Information Systems - Intelligent Information Processing Systems, Natural Language Processing, Affective Computing and Artificial Intelligence, and an Attempt to Build a Conversational Nursing Robot

Edited by Kazuyuki Matsumoto

Published in London, United Kingdom

IntechOpen

Supporting open minds since 2005

Information Systems - Intelligent Information Processing Systems, Natural Language Processing, Affective Computing and Artificial Intelligence, and an Attempt to Build a Conversational Nursing Robot
http://dx.doi.org/10.5772/intechopen.87694
Edited by Kazuyuki Matsumoto

Contributors
Kazuyuki Matsumoto, Seiji Tsuchiya, sarra abidi, Myriam Fakhri, Fathia Bettaher, Yuko Yasuhara, Kyoko Osaka, Ryuichi Tanioka, Feni Betriana, Tetsuya Tanioka, Yoshihiro Kai, Rozzano C. Locsin, Hirokazu Ito, Michael Joseph S. Diño, Irvin L. Ong, Yuki Oobayashi, Misao Miyagawa, Rozzano Locsin, Minoru Yoshida, Kenji Kita

First published in London, United Kingdom, 2021 by IntechOpen
IntechOpen is the global imprint of INTECHOPEN LIMITED, registered in England and Wales, registration number: 11086078, 5 Princes Gate Court, London, SW7 2QJ, United Kingdom
Printed in Croatia

British Library Cataloguing-in-Publication Data
A catalogue record for this book is available from the British Library

Additional hard and PDF copies can be obtained from orders@intechopen.com

Information Systems - Intelligent Information Processing Systems, Natural Language Processing, Affective Computing and Artificial Intelligence, and an Attempt to Build a Conversational Nursing Robot
Edited by Kazuyuki Matsumoto
p. cm.
Print ISBN 978-1-83880-853-2
Online ISBN 978-1-83962-360-8
eBook (PDF) ISBN 978-1-83962-361-5

For EU product safety concerns:
IN TECH d.o.o., Prolaz Marije Krucifikse Kozulić 3, 51000 Rijeka, Croatia, info@intechopen.com or visit our website at intechopen.com.

Meet the editor

Dr. Kazuyuki Matsumoto received a Ph.D. in 2008 from the Faculty of Engineering, Tokushima University, Japan, where he is currently an associate professor. His research interests include affective computing, artificial intelligence, Kansei robotics, and natural language processing. He is a member of the Association for the Advancement of Affective Computing (AAAC), Association for Natural Language Processing (ANLP), Institute of Electrical Engineers of Japan (IEEJ), Japan Society of Kansei Engineering (JSKE), Japanese Society for Artificial Intelligence (JSAI), Institute of Electronics, Information and Communication Engineers (IEICE), Human Interface Society (HIS), and Information Processing Society of Japan (IPSJ).

Contents

Preface

This book is a collection of chapters describing research examples and empirical studies on recent intelligent information processing systems and their applications, especially nursing robots, which will become increasingly important in the future. It can be said that the intelligent information processing systems envisioned in artificial intelligence (AI) research aim to "process things more intelligently, smartly, and humanely," a step forward from the information processing systems'goal of "processing things quickly and reliably." In addition, "human-like" processing also means making the mistakes humans tend to make, so there is room for debate about how human-like it should be. Knowledge information processing, especially information retrieval, personalization in the recommendation, intelligent dialogue processing, emotional information processing, and so on, are topics that many researchers have been working on using various approaches. These have influenced the recent rapid development of AI technology. One urgent task is to combine these mature information processing technologies and realize interactive nursing care robots to solve the shortage of caregivers caused by the declining birthrate and aging population, which has become a social problem.

The first section contains two chapters on numerical mining techniques and personalization in information processing systems.

Chapter 1 is a survey of recent mining literature on how to handle numerical information in text mining techniques. Although many methods have been proposed for retrieving textual information, retrieval based on numerical values that consider units, scales, and so on, has not been conducted in conventional text mining research for language. While it is necessary to organize and store numerical values in a database, tabular data on the Web are not always described in a uniform format, and it is necessary to combine text mining techniques.

Chapter 2 describes a research case study of a web service system that makes personal adaptations according to user interactions. To obtain the necessary information on the Web, the user's own characteristics must be fully considered. In this context, a service composition approach that enables dynamic service composition aimed at satisfying user needs based on ontology and user profile information, and a personalization approach that enables service reuse based on user context, are proposed.

The second section contains two chapters on techniques essential to developing AI, such as associative techniques, common sense judgment, and emotional information processing.

Chapter 3 describes the usefulness of associative knowledge techniques, which will be important in realizing the next generation of AI.

Chapter 4 presents experiments with real data on a method that integrates machine learning using corpus-based examples and syntactic knowledge using deep learning techniques. The results show that lexical knowledge is also important to compensate for the insufficient amount of data.

The third section contains two chapters on robotics research in nursing care.

Chapter 5 explores the challenges of developing humanoid robot conversational dialogues for nursing, especially caregiving, and discusses the introduction of robots into clinical practice. It identifies the main issue that needs improvement, that of the robot's speech (intonation, vocal range, speech rate, etc.).

Chapter 6 is a survey of the effectiveness of robot therapy in improving symptoms of dementia patients and a case study by the authors.

The fourth section contains chapters that examine ways to realize the intelligence that will need to be provided to humanoid nursing robots and the potential ethical dilemma of introducing humanoid nursing robots into the medical field from the perspective of nurses.

Chapter 7 describes the nursing situation in Japan, PsyNACS as a specialized nursing database, future nursing robots, and an evolving artificial brain that links PsyNACS with AI using deep learning and natural language processing (NLP). Chapter 8 discusses the ethical dilemma of introducing robots into nursing from multiple perspectives and, with regards to the process of developing robots, states the importance of discussion and collaboration with interdisciplinary teams to protect patient rights and maintain safety.

Kazuyuki Matsumoto
Tokushima University,
Graduate School of Technology, Industrial and Social Sciences,
Tokushima, Japan

Section 1

Text Mining Methods and Personalized Web Services

Chapter 1

Mining Numbers in Text: A Survey

Minoru Yoshida and Kenji Kita

Abstract

Both words and numerals are tokens found in almost all documents but they have different properties. However, relatively little attention has been paid in numerals found in texts and many systems treated the numbers found in the document in ad-hoc ways, such as regarded them as mere strings in the same way as words, normalized them to zeros, or simply ignored them. Recent growth of natural language processing (NLP) research areas has change this situations and more and more attentions have been paid to the numeracy in documents. In this survey, we provide a quick overview of the history and recent advances of the research of mining such relations between numerals and words found in text data.

Keywords: text mining, numeracy, survey, embedding, natural language inference

1. Introduction

Natural language processing (NLP) is a research field to make machines understand the meaning of a text data, which is a typically a list of words. In some cases, texts are not understandable in their closed form, i.e., without understanding the data other than the words. Numerals are an important form of data in such non-word data not only because many documents are accompanied with related metadata such as publish dates expressed in the form of numbers, but also because the document themselves contain numerals such as "three people", "500 dollars", and "90 cm."

Jointly mining texts and their associated numerical metadata has many variations and many studies have been proposed. For example, predicting stars given with product review texts is typical task of such research areas. Location-aware text mining can be considered as mining of association rules between words and positional data (i.e., longitude and latitude). Even joint learning of texts and images can be seen as mining relations between texts and associated RGB data.

In contrast to such *grounding*-type research, studies on mining numerals explicitly written in text have been getting little attention. However, recently more and more research studies are proposed on this area partly due to the recent advance of deep neural network-based language modeling.

In this survey, we try to provide a quick overview of the history and recent advances of this research field ranging from traditional tasks like information retrieval to emerging ones such as numerical reading comprehension.

2. Traditional tasks

Firstly, we give a survey on the systems which consider treatment of numbers for the traditional tasks such as information retrieval (IR), question answering (QA), and information extraction (IE). Some of the questions or queries of these tasks require the answers to be numbers, hence requiring appropriate treatment of numbers found in the target text.

2.1 Question answering

Question Answering (QA) is a task to find appropriate answers from text to the questions given also in text. Because many type of questions require the answers to be numbers e.g., 8,848 (8,849) (meters) is the answer of the question "how tall is Everest?", some existing QA systems treat numbers appropriately, typically in ad-hoc heuristical ways.

For example, IBM's PIQUANT system for TREC2003 [1] have had the sanity checking module, which use the Cyc knowledge base to check the given answer is valid intervals found in Cyc, e.g., rejecting "200 miles" for the questions for height of a mountain, having the knowledge "mountains are between 1,000 and 30,000 high" from Cyc. Moriceau [2] consider more complicated situation where several numeric answers can be extracted from different Web pages in QA system. They proposed a way to integrate considering the nature of numbers such as number approximations.

2.2 Information retrieval

Similarly to QA tasks, some Information Retrieval (IR) systems return the direct "answers" to the query. Therefore, appropriate treatment of numbers is required for some type of queries. For example, Banerjee et al. [3] introduced Quantity Consensus Queries (QCQs), the answers for which is the quantity intervals, such as "driving time from Paris to Nice". Their proposed algorithm propose and rank intervals considering whether returned snippets is included in the intervals or not. Sarawagi and Chakrabarti [4] proposed a system to answer quantity queries on Web tables such as "escape velocity jupiter." Their system contain the modules to interpret the numbers presented in the table cells to improve the accuracy.

On the contrary, queries also can be numbers. Yoshida et al. [5] proposed a suffix array-based text mining system enhanced with treatment of numbers, which accept range queries like "[1,000 - 10,000] ft'.

2.3 Information extraction

Information extraction (IE) is another type of systems that return the answers to the questions, but in this case the questions are given a priori such as "extract all dates and places of events found in the given documents." Many extracted information is in numerals, so special treatment of numbers often contributes to the improvement of the performance of IE systems.

For example, Bakalov and Fuxman [6] proposed a system to extract numerical attributes of objects given attribute names, seed entities, and related Web pages and properly distinguish the attributes having similar values.

Table 1 summarizes these systems.

System	Special Treatment for Numerals
Question Answering Systems	
IBM's PIQUANT [1]	Matching numerals with values in knowledge base
Moriceau [2]	Integrating multiple numerals
Information Retrieval Systems	
Banerjee et al. [3]	Number interval estimation for a collection of numerals
Sarawagi and Chakrabarti [4]	Understanding numerals in tables
Yoshida et al. [5]	Number range queries for text mining
Information Extraction Systems	
Bakalov and Fuxman [6]	Distinguish similar but different numerals

Table 1.
Systems for Traditional Tasks Considering Numerals.

3. Numerical common sense acquisition

Numerical common senses acquisition is a task to obtain numerical common senses, e.g., the height of mountains have a typical values "1,000 - 10,000 meters". Many numerals found on text typically describe some *attributes* of *objects* such as "25 C" for the temperature of some city, "170 cm" for the height of some person, etc. Obtaining such numerical common senses can contribute to improving various kinds of systems, e.g., anomaly detection or dialogue systems, etc.

We introduce two type of tasks in this type of research. One is a task to directly extract the common senses, and the other is a task to acquire such knowledge as language model parameters.

3.1 Pattern-based extraction of numerical common senses

In this task, the input is a large collection of texts.[1] The output is a database for "typical values" of something.[2]

Typical methods for this task is to use pattern matching to obtain numerals for each attribute described in the given text. For example, the value "80" can be extracted from the sentence "The size of the dog is 80 cm." using the pattern "the size of the A is # cm."

3.1.1 Previous methods

Aramaki et al. [7] proposed to obtain physical size of entities by using Web search with patterns like "book (*cm x *cm)". Bagherinezhad et al. [8] proposed to use knowledge obtained using these patterns with object detection from images to achieve more reliable object size knowledge. Davidov and Rappoport [9] proposed similar approach but augment their method by obtaining terms similar to given object using the Web and WordNet. Takamura and Tsujii [10] took similar approach by using Web search for linguistic patterns e.g., "the size of A", but they enhanced their patterns with more indirect clues such as WordNet relations,

[1] It includes the case where the system uses Web search engines where the huge amount of texts are behind the search engines.

[2] It is an attribute of object in most cases.

n-gram corpus for the explicit patterns, e.g., "A is longer than B", and implicit patterns, e.g., "put A in B", through a machine learning approach to determine their weights.

Narisawa et al. [11] proposed to obtain numerical common sense by searching numerical expressions in Web corpus, and calculating distribution of numbers given contexts that are given syntactically such as "verb=give, subj=he, ..." and predict labels for given numbers in text, such as *small*, *normal*, *large*.

Recently, a large dataset called Distribution over Quantities (DoQ), was provided by Elazar et al. [12]. It contains ten dimensions (TIME, CURRENCY, LENGTH, AREA, VOLUME, MASS, TEMPERATURE, DURATION SPEED, VOLTAGE) for various kinds of words including nouns, adjectives, and verbs. They explored co-occurrence of words and numeracy in large Web data.

Table 2 summarizes these approaches and **Table 3** shows the existing data sets.

3.2 Prediction of numbers in sentences

Some researchers tried to acquire numerical common senses as parameters of language modeling. In this type of research, the system directly predicts numbers to fill in the blanks in texts, or assessing feasibility of the number presented in text, without explicitly collecting above-mentioned knowledge bases.

3.2.1 Task definition

In this task, the input is a sentence, or document, where the position for a numeral is masked. The system then outputs a likely value for the masked position. For example, given the sentence "my five-year-old son is [MASK] cm tall.", the

System	Example Patterns	Source
Aramaki et al. [7]	"A (*cm x *cm)"	Web search
Bagherinezhad et al. [8]	"A * x * cm"	Web search, Flickr
	"A is * cm tall"	
	Objects in images	
Davidov and Rappoport [9]	"A is * cm tall"	Web Search, TREC
Takamura and Tsujii [10]	"the size of A",	Web search
	"A is longer than B",	
	WordNet	
Narisawa et al. [11]	Syntactic Patterns	Web corpus
Elazar et al. [12]	co-occurrence	Web corpus

Table 2.
Systems for Numerical Common Sense Acquisition.

Data	Source	Method
DoQ [12]	Web corpus	Collecting co-occurred numerals for each word

Table 3.
Data set for Numerical Common Sense Acquisition.

system is required to answer the likely value to be filled in the position of "[MASK]."

Because the input is a sequence of words, encoder-decoder models are applicable to this task. Especially, the BERT language model is a good match for this problem. BERT is a deep neural network model that consists of modules called *Transformers*. It is trained on the task where the input is a sequence of words with special "[MASK]" tokens, and one of the output is the estimated original word for the position of "[MASK]".

3.2.2 Previous approaches

Several BERT models pretrained on a huge size of text data are available to the public. Using such pretrained language model to predict or assess the numeracy in documents is an emerging trend. Typically, the models are enhanced with the ability to predict numbers by, simply using masked language models by replace the numbers to be predicted with [MASK] tokens, or by adding numeracy inference modules into language models or by fine-tuning setting where output is a *discretized* versions of target numeracy.

Zhang et al. [13] investigated how pretrained language model like BERT can predict (the *discretized* version of) the attribute with continuous numeric values such as MASS or PRICE with evaluation with DoQ. Chen et al. [14] proposed a task of predicting the magnitude of hidden numerals in text and provided a large dataset called Numeracy-600 K. They also reported CNN and RNN-based models to solve this task. Berg-Kirpatrick and Spokoyny [15] proposed more advanced model using BERT and reported that using BERT was better than other models including BiGRU.

On the other hand, Lin et al. [16] considered more difficult task to predict *accurate* number to be filled in the blank in text, like "A bird usually has [MASK] legs". They reported that the current pretrained models including BERT and RoBERTa performed poorly.

A language model that did not use encoder-decoder model was also proposed. Spithourakis and Riedel [17] proposed a language model for a sequence of words and numerals, which gives the probability for words and numerals simultaneously. For example, it gives the probability of the numeral "50,000" appearing just after the word sequence "the number of video-game consoles I have is". They introduced the probabilities of being words or numerals for each token, and modeled the probability for numerals independently of that of words, using some variations including digit-based RNN and mixture of Gaussians.

Table 4 summarizes the approaches and **Table 5** shows the dataset for this task.

System	Model	Task
Zhang et al. [13]	BERT fine tuning	Magnitude[a] prediction
Chen et al. [14]	CNN, RNN	Magnitude prediction
Berg-Kirpatrick and Spokoyny [15]	BERT	Magnitude prediction
Lin et al. [16]	BERT, RoBERTa	Accurate number prediction
Spithourakis and Riedel [17]	generative probability	Perplexity maximization
	(digit-based RNN and Gaussian)	

[a]*Discretized (binned) values, e.g., categorized by the number of digits of numerals.*

Table 4.
Number Prediction Data Sets.

Data	Source	Task
Numeracy-600 K [14]	Market comments from Reuters	Magnitude prediction

Table 5.
Number Prediction DataSet.

4. Numeracy embeddings

Embedding or *distributed representation* of words has become basic building blocks for natural language processing in recent years. It represents each word by high-dimension vectors (typically with 50 dimensions or more) of real values. These vectors reflect the meaning of words, such as words with similar meaning are represented by similar[3] vectors. Some researchers have investigated how numeracy itself is modeled in such pre-trained word embedding vectors.

4.1 Task definition

Embedding vectors are also assigned to numerals such as "three", "100", "million", etc. Popular word embeddings like word2vec do not distinguish these numerals from other words, i.e., the learning algorithms for these vectors treat numerals and other words equally. So, it is not obvious these word vectors appropriately reflect the meaning of numbers, such as "100 is larger than 3" and "4 is the next number of 3", etc. *Numeracy embedding* is a task to embed such numerals in appropriate vector representation.

4.2 Investigating pre-trained word vectors

Nowadays, word embeddings learned using huge size of corpus are provided by various researchers. Some researchers investigated how or whether these pretrained word vectors appropriately represent numerals.

Naik et al. [18] used GloVe, FastText, and SkipGram vectors. They compares similarity of embedding vectors for numbers. They used two types of tasks: one is for magnitude, e.g., vector for 4 should be more similar to 3 than 1000000, and the other is for numeration, e.g., vector for *three* should be more similar to 3 than *billion*. Contextualized word vectors were also considered. Wallace et al. [19] found that the pretrained language models for DROP, which is numeracy entailment task mentioned later sections, already captures numeracy, by testing if BiLSTM model with pre-trained embedding pass some tests such as list maximum, decoding (e.g., convert the string "five" to 5), taking a sum of two numbers.

4.3 Obtaining word vectors for numerals

On the other hand, developing algorithms specialized to obtain word vectors for numerals beyond pre-trained word vectors have been proposed by some researches in recent years.

Jiang et al. [20] proposed to obtain embedding for numbers by directly applying Skip-Gram models to obtain embeddings for numbers taking into consideration of meaning of numbers by taking weighted average of embeddings which is

[3] Similarity of vectors is typically defined by inner product or cosine similarity of vectors.

System	Vector	Algorithm	Task
Naik et al. [18]	Glove, FastText,	Cosine Similarity	Magnitude[a] prediction,
	SkipGram		Numeration[b] prediction
Wallace et al. [19]	NAQANET[c]	biLSTM	List maximum, Decoding, Sum
Jiang et al. [20]	Learned	Cosine Similarity	Magnitude prediction,
			Numeration prediction
			Numeral prediction
Sundararaman et al. [21]	Learned	Cosine Similarity,	Magnitude prediction,
		BiLSTM	Numeration prediction
			List maximum, Decoding, Sum

[a]*e.g., 4 is more similar to 3 than 1000000.*
[b]*e.g., 'three' is more similar to 3 than 'billion'.*
[c]*A system proposed for DROP dataset.*

Table 6.
Numeracy Embedding Systems.

numerically similar to the target number. They find "prototype numbers" by clustering, and represent numbers as a weighted average of these prototypes. Sundararaman et al. [21] proposed to learn embeddings for numbers, which reflect the distance of two numbers in the number line, independently from words.

Table 6 summarizes these previous methods.

5. Numerical reading comprehension and numerical textural entailment detection

More complex tasks such as textual entailment detection or reading comprehension also require treatment of numbers appropriately to answer some of the questions. We first mention on some early works for these tasks and then introduce some recent systems.

5.1 Task definition

Textual entailment detection is a task to find, given some texts, sentences which are true if the given text (called *hypothesis*) are true. The situation become more complicated if the sentences contains numerals because it requires numerical knowledge to understand the meaning of sentence. For example, we can say that the sentence "five people are in the house." is true given the hypothesis "two men and three women are in the house.", but it requires mathematical knowledge that two plus three equals five.

Some early works for this task include numeracy modules. The system by Tsuboi et al. [22] for textual entailment recognition task (RITE) in NTCIR-9 consider temporal expression matching such as "the first half of Nth century" to the appropriate interval. The system by Iftene and Moruz [23] implemented the special rules for numbers which create intervals considering expressions like "more than" or "over" for the Recognizing Textual Entailment (RTE-6) task.

Reading comprehension is a more complicated task, where the system is required to answer various types of questions.

5.2 Numeracy-focused data sets

Aforementioned studies mainly focused on the "range" of the numbers, i.e., they simply treat numbers as points or distributions defined on the number line. However, reading comprehension tasks require more advanced numeric skills such as addition, average, maximum, etc., into language models.

This line of research typically constructs the dataset for numeracy understanding task by selecting numeracy-related data from existing datasets for reading comprehension, natural language inference, or entailment. The selected data contain many questions that require understanding and calculation on numbers beyond simple range- or distribution-based treatment of numbers.

Roy et al. [24] proposed the task of Quantity Entailment, which require numeric reasoning to answer. Their dataset included the corpus from datasets for Recognizing Textual Entailment (RTE) task. They also proposed a method to solve these problems with CRF-based recognition of quantity part of the text, and rule-based recognition of entailment.

Ravichander et al. [25] proposed the EQUATE framework for quantitative reasoning in textual entailment, such as determining "5855 of lambs are back" is correct given the premise "6048 lambs is either black or white and there are 193 white ones." DROP proposed by Dua et al. [26] require systems to do operations such as addition, counting, or sorting. The type of questions and answers in DROP dataset varies widely, such as the question "Where did Charles travel to first" given passages "In 1517, the King sailed to Castle. ... In 1518, he traveled to Barcelona." State-of-the-art methods for reading comprehension performed poorly for these datasets (both of EQUATE and DROP) and the authors concluded that more advanced methods are required for these new tasks for numeric reading comprehension.

Table 7 summarizes these datasets.

5.3 Methods

Given these datasets, more advanced models for them have been proposed. Typical approaches given the recent advance of deep neural network technologies is to use sequence-to-sequence (seq2seq) model for this task. In seq2seq models, the sequence of words can be feed as input directly to the system, then the system also returns another sequence of words as the output. Especially, recent pretrained language models including BERT already contains language models trained on huge amount of text documents, and they can be trained to return appropriate word sequence by being trained on relatively small set of training samples (i.e., the pair of input documents and "correct" or appropriate output for each input.) of a given task.

Rozen et al. [27] reported that performance for existing natural language inference (NLI) datasets can be improved by augmenting the dataset with synthetic adversarial datasets including the ones generated by rule-based replacement of numeric expressions found in the dataset. Geva et al. [28] reported that adding

Data Set	Source	Task
DROP [26]	Wikipedia	Question answering
EQUATE [25]	News, Reddit, Synthetic	Entailment

Table 7.
Numeracy-Focused Data Sets.

System	Method	Base Model
Rozen et al. [27]	Data augmentation	BERT
Geva et al. [28]	Subtask addition	BERT
Ran et al. [29]	Adding reasoning module	NAQANet

Table 8.
Numeracal Reading Comprehension Systems.

synthetic numerical tasks to BERT pretraining steps with fine tuning on DROP dramatically improved the score for DROP. Ran et al. [29] proposed to inject graph-based numerical reasoning module between embedding and prediction modules, which outperformed existing machine reading comprehension models on the DROP dataset.

Table 8 summarizes these approaches.

6. Solving math word problems

Math word problem texts are a typical type of documents that contain numerals and words extensively and require deep understanding of the meaning of numerals. Developing a system that automatically solve math word problems is thus one major research task in this area.

6.1 Task definition

In this task, the problem is given in a text that contains numerals, e.g., "How much How much would it cost to buy 12 apples at 1.1 dollars each?", and systems are required to provide a solution for the problem, e.g., $12 \times 1.1 = 13.2$ dollars. Recent approaches for this task typically use deep neural networks that take a sequence of words as inputs. These inputs are transformed through several layers and used to produce the final output. Variety of output forms are considered by previous methods, including simple seq2seq models (i.e., outputs are also sequences of words) and sequence-to-tree models (i.e., outputs are in tree forms that represent equations to calculate the answers.)

Sequence-to-sequence (seq2seq) is a typical approach for this task. Ling et al. [30] provided their original dataset with 100,000 samples, and proposed a method to generate *answer rationales* which are human-readable instructions to derive the answers using a sequence-to-sequence (seq2seq) model. Saxton et al. [31] investigate the ability of existing sequence to sequence architectures including Transformer for mathematical reasoning (e.g., "Solve 41 + 132") with free-form texts.

Some researchers have tried to produce graphs that represent the mathematical operations to directly produce the answers to the questions. Amini et al. [32] provided a dataset for math word problems called MathQA. They also proposed the sequence-to-Program model to solve this task. The approach by Zhang et al. [33] uses a new architecture called Graph2Tree, which uses graphs constructed from texts independently from BiLSTM encoders. They tested their system on MAWPS data set. [34] Lample and Charton [35] showed that neural models can solve mathematical problems such as symbolic integration and solving differential equations using sequence-to-sequence approaches.

Table 9 summarizes the existing data sets and **Table 10** summarizes the systems for this task proposed so far.

System	Method	Input	Output
Ling et al. [30]	Seq2seq	GRE/GMAT-style	Answer rationales[a]
Saxton et al. [31]	Seq2seq	Free-form text	Solution
Amini et al. [32]	Sequence-to-Program	GRE/GMAT-style	Solution
Zhang et al. [33]	Seq2seq, Graph2Tree	Free-form text	Solution
Lample and Charton [35]	Seq2seq	Equations	Solution

[a]*linguistic instructions to calculate answers.*

Table 9.
Systems for Math Word Problem Solving.

Data	Source
MAWPS [34]	Collected through the Web interface
AQuA [30]	Originally created
MathQA [32]	Enhancing AQuA

Table 10.
Math Word Problem Datasets.

7. Other tasks

Yoshida et al. [36] considered a problem of estimating appropriate units for the numbers found in Wikipedia tables when units were omitted. Elazar and Goldberg [37] considered the problem to infer the omitted head related to numerals such as "It is worth about two million __."

Chen et al. [38] proposed the numeral attachment task, which determine what entity is the number presented in text related. They also proposed the task of numeral categorization, which is to classify numerals presented in financial text into 7 or 17 categories [39].

The task proposed by Chaganty and Liang [40] was to describe given numerals by examples, such as "$131 million is about the cost to employ everyone in Texas over a lunch period."

8. Conclusions

The relations between numerals and words found in text data has been paid little attention compared to other areas in natural language processing. This paper provided the overview of this field ranging from the systems for traditional tasks such as information retrieval tor the relatively recent tasks like reading comprehension.

We categorized the previous researches into 6 types: traditional tasks, numerical common sense acquisition, numeracy embeddings, numerical reading comprehension, solving math word problems, and others. The first two tasks have been studied relatively long time, while the remaining topics is emerging with recent advances of neural language models.

In Section 2, we introduced some previous systems that have numerical modules for traditional tasks like QA, IE, and IR. In Section 3, we introduced numerical common sense acquisition where typical approaches are pattern-based extraction and parameter estimation for language models. In Section 4, numeracy embedding, where the goal is assigning appropriate real-valued vectors to numerals, was introduced, Section 5 introduced numerical reading comprehension and numerical entailment, that require more advanced numerical understanding of text. The task of solving math word problems, which are typical type of texts that contain numerals extensively, was introduced in Section 6, and Section 7 touched on other unique tasks.

Recent increase of the dataset and resources focusing on numeracy will accelerate the development of the systems with the ability of understanding numeracy in text.

Acknowledgements

This work was supported by JSPS KAKENHI Grant Numbers JP21K12141 and JP20K12027.

Author details

Minoru Yoshida* and Kenji Kita
Tokushima University, Tokushima, Japan

*Address all correspondence to: mino@is.tokushima-u.ac.jp

References

[1] John M. Prager, Jennifer Chu-Carroll, Krzysztof Czuba, Christopher A. Welty, Abraham Ittycheriah, Ruchi Mahindru: IBM's PIQUANT in TREC2003. TREC 2003: 283-292

[2] Numerical data integration for cooperative question-answering Véronique Moriceau Proceedings of the Workshop KRAQ'06: Knowledge and Reasoning for Language

[3] Somnath Banerjee, Soumen Chakrabarti, Ganesh Ramakrishnan: Learning to rank for quantity consensus queries. SIGIR 2009: 243-250

[4] Sunita Sarawagi, Soumen Chakrabarti: Open-domain quantity queries on web tables: annotation, response, and consensus models. KDD 2014: 711-720

[5] Minoru Yoshida, Issei Sato, Hiroshi Nakagawa, Akira Terada: Mining Numbers in Text Using Suffix Arrays and Clustering Based on Dirichlet Process Mixture Models. PAKDD (2) 2010: 230-237

[6] Anton Bakalov, Ariel Fuxman, Partha Pratim Talukdar, Soumen Chakrabarti: SCAD: collective discovery of attribute values. WWW 2011: 447-456

[7] Eiji Aramaki, Takeshi Imai, Kengo Miyo, Kazuhiko Ohe: UTH: SVM-based Semantic Relation Classification using Physical Sizes. SemEval@ACL 2007: 464-467

[8] Hessam Bagherinezhad, Hannaneh Hajishirzi, Yejin Choi, Ali Farhadi: Are Elephants Bigger than Butterflies? Reasoning about Sizes of Objects. AAAI 2016: 3449-3456

[9] Dmitry Davidov, Ari Rappoport: Extraction and Approximation of Numerical Attributes from the Web. ACL 2010: 1308-1317

[10] Hiroya Takamura, Jun'ichi Tsujii: Estimating Numerical Attributes by Bringing Together Fragmentary Clues. HLT-NAACL 2015: 1305-1310

[11] Katsuma Narisawa, Yotaro Watanabe, Junta Mizuno, Naoaki Okazaki, Kentaro Inui: Is a 204 cm Man Tall or Small ? Acquisition of Numerical Common Sense from the Web. ACL (1) 2013: 382-391

[12] Yanai Elazar, Abhijit Mahabal, Deepak Ramachandran, Tania Bedrax-Weiss, Dan Roth: How Large Are Lions? Inducing Distributions over Quantitative Attributes. ACL (1) 2019: 3973-3983

[13] Xikun Zhang, Deepak Ramachandran, Ian Tenney, Yanai Elazar, Dan Roth: Do Language Embeddings capture Scales? EMNLP (Findings) 2020: 4889-4896

[14] Chung-Chi Chen, Hen-Hsen Huang, Hiroya Takamura, Hsin-Hsi Chen: Numeracy-600K: Learning Numeracy for Detecting Exaggerated Information in Market Comments. ACL (1) 2019: 6307-6313

[15] Taylor Berg-Kirkpatrick, Daniel Spokoyny: An Empirical Investigation of Contextualized Number Prediction. EMNLP (1) 2020: 4754-4764

[16] Bill Yuchen Lin, Seyeon Lee, Rahul Khanna, Xiang Ren: Birds have four legs?! NumerSense: Probing Numerical Commonsense Knowledge of Pre-Trained Language Models. EMNLP (1) 2020: 6862-6868

[17] Georgios P. Spithourakis, Sebastian Riedel: Numeracy for Language Models: Evaluating and Improving their Ability to Predict Numbers. ACL (1) 2018: 2104-2115

[18] Aakanksha Naik, Abhilasha Ravichander, Carolyn Penstein Rosé,

Eduard H. Hovy: Exploring Numeracy in Word Embeddings. ACL (1) 2019: 3374-3380

[19] Eric Wallace, Yizhong Wang, Sujian Li, Sameer Singh, Matt Gardner: Do NLP Models Know Numbers? Probing Numeracy in Embeddings. EMNLP/IJCNLP (1) 2019: 5306-5314

[20] Chengyue Jiang, Zhonglin Nian, Kaihao Guo, Shanbo Chu, Yinggong Zhao, Libin Shen, Kewei Tu: Learning Numeral Embedding. EMNLP (Findings) 2020: 2586-2599

[21] Dhanasekar Sundararaman, Shijing Si, Vivek Subramanian, Guoyin Wang, Devamanyu Hazarika, Lawrence Carin: Methods for Numeracy-Preserving Word Embeddings. EMNLP (1) 2020: 4742-4753

[22] Yuta Tsuboi, Hiroshi Kanayama, Masaki Ohno, Yuya Unno: Syntactic Difference Based Approach for NTCIR-9 RITE Task. NTCIR 2011

[23] Adrian Iftene, Mihai Alex Moruz: UAIC Participation at RTE-6. TAC 2010

[24] Subhro Roy, Tim Vieira, Dan Roth: Reasoning about Quantities in Natural Language. Trans. Assoc. Comput. Linguistics 3: 1-13 (2015)

[25] Abhilasha Ravichander, Aakanksha Naik, Carolyn Penstein Rosé, Eduard H. Hovy: EQUATE: A Benchmark Evaluation Framework for Quantitative Reasoning in Natural Language Inference. CoNLL 2019: 349-361

[26] Dheeru Dua, Yizhong Wang, Pradeep Dasigi, Gabriel Stanovsky, Sameer Singh, Matt Gardner: DROP: A Reading Comprehension Benchmark Requiring Discrete Reasoning Over Paragraphs. NAACL-HLT (1) 2019: 2368-2378

[27] Ohad Rozen, Vered Shwartz, Roee Aharoni, Ido Dagan: Diversify Your Datasets: Analyzing Generalization via Controlled Variance in Adversarial Datasets. CoNLL 2019: 196-205

[28] Mor Geva, Ankit Gupta, Jonathan Berant: Injecting Numerical Reasoning Skills into Language Models. ACL 2020: 946-958

[29] Qiu Ran, Yankai Lin, Peng Li, Jie Zhou, Zhiyuan Liu: NumNet: Machine Reading Comprehension with Numerical Reasoning. EMNLP/IJCNLP (1) 2019: 2474-2484

[30] Wang Ling, Dani Yogatama, Chris Dyer, Phil Blunsom: Program Induction by Rationale Generation: Learning to Solve and Explain Algebraic Word Problems. ACL (1) 2017: 158-167

[31] David Saxton, Edward Grefenstette, Felix Hill, Pushmeet Kohli: Analysing Mathematical Reasoning Abilities of Neural Models. ICLR (Poster) 2019

[32] Aida Amini, Saadia Gabriel, Shanchuan Lin, Rik Koncel-Kedziorski, Yejin Choi, Hannaneh Hajishirzi: MathQA: Towards Interpretable Math Word Problem Solving with Operation-Based Formalisms. NAACL-HLT (1) 2019: 2357-2367

[33] Jipeng Zhang, Lei Wang, Roy Ka-Wei Lee, Yi Bin, Yan Wang, Jie Shao, Ee-Peng Lim: Graph-to-Tree Learning for Solving Math Word Problems. ACL 2020: 3928-3937

[34] Rik Koncel-Kedziorski, Subhro Roy, Aida Amini, Nate Kushman, Hannaneh Hajishirzi: MAWPS: A Math Word Problem Repository. HLT-NAACL 2016: 1152-1157

[35] Guillaume Lample, François Charton: Deep Learning For Symbolic Mathematics. ICLR 2020

[36] Minoru Yoshida, Kazuyuki Matsumoto, Kenji Kita: Table Topic

Models for Hidden Unit Estimation. AIRS 2016: 302-307

[37] Yanai Elazar, Yoav Goldberg: Where's My Head? Definition, Dataset and Models for Numeric Fused-Heads Identification and Resolution. Trans. Assoc. Comput. Linguistics 7: 519-535 (2019)

[38] Chung-Chi Chen, Hen-Hsen Huang, Hsin-Hsi Chen: Numeral Attachment with Auxiliary Tasks. SIGIR 2019: 1161-1164

[39] Chung-Chi Chen, Hen-Hsen Huang, Yow-Ting Shiue, Hsin-Hsi Chen: Numeral Understanding in Financial Tweets for Fine-Grained Crowd-Based Forecasting. WI 2018: 136-143

[40] Arun Tejasvi Chaganty, Percy Liang: How Much is 131 Million Dollars? Putting Numbers in Perspective with Compositional Descriptions. ACL (1) 2016

Chapter 2

Towards a Personalized Web Services Composition Approach

Sarra Abidi, Fathia Bettaher and Myriam Fakhri

Abstract

Generally available Web Services (WS) can not meet the complex needs of users and their adaptation to the environment remains a major problem for the design of information systems. The web services composition comes to address the satisfaction of new and complex needs such as the process we find in most organizations. Its purpose is to perform several services to meet user demand. **The satisfaction of a user needs a dynamic and reusable environment to meet those needs.** In this context, the user interactions are essential. From there, in this work, we define two objectives: **i)** propose a service composition approach that allows dynamic services composition and its purpose is to meet a need. **ii)** Propose a personalization approach for Web services composition which allows the reuse of services while adopting for the context of each user. Our approach is based on the use of ontologies and user profile.

Keywords: Web services composition, personalization, ontologies, user profile

1. Introduction

Nowadays, we find a large number of services available on the Web and in different directories, where a Web service is an application made available on internet. These services are generally defined by their function, input/output [1–3], which allow their reuse. However, user requirements are continually evolving, so the available services can not meet all needs especially the most complex ones. The composition of services is coming precisely to answer these two questions. After analyzing several definitions (Fekih et al) [4], (Shanchen et al) [5], (Yuan et al) [6], we hold two views on the services composition; According to (Shanchen et al) [5] who has a vision process on the services composition: "The composition is the selection process, combination and implementation services to accomplish a given objective."

A second view more global is that of (Fekih et al) [4], "The composition then being an effective way to create, run, and maintain services that depend on other services."

Based on these definitions, we believe that the services composition has essentially two objectives:

- Combine basic services to meet a need that no service has been solved.
- Define the interaction between services.

We distinguish two types of composition services; "Orchestration" is the process of programming a central engine which, on its part, controls and calls all services according to a predefined process. Added to that, it defines the order of execution of services [7]. "Choreography", for its part, aims to achieve a common goal between a set of Web services. The collaboration between each web service collection (part of the composition) is described by control flow [8].

Regarding categories of the services composition, we distinguish between; on the one hand, a static composition which uses a fixed manner services defined in advance, which are unchanged and independent of the client context [9] and "a dynamic composition" which occurs when running within the constraints required by the client [9]. "A semi-dynamic composition" combines the two types mentioned previously. On the other hand, we find a "manual composition" which considers that the responsible is the user who generates the composition by hand via a text editor and without using dedicated tools. "The semi-automatic composition" is a step forward compared to the static composition, to the extent that its techniques make semantic proposals to help in the selection of Web services. "Automatic composition" is the automation of the entire composition process, without any user intervention.

Given the continuous increase of heterogeneous information sources and the diversity of user requirements, retrieval information systems deliver massive results. In this context, these results generate subsequent information that disorients the user to distinguish what is relevant from what is not.

In literature, the term "personalization" knows a success. Let us look at the opinion of (Kostadinov) [10], which announces that "the personalization of information comes from a set of individual preferences, by ordering criteria or semantic rules". Such specifications allow obtaining the quality level desired and data arrangements. In this context, personalization of information is a major challenge for the IT industry to the extent that the relevance of the information delivered its intelligibility and its adaptation to the uses and preferences constitute as well as key factors of success or rejection of these systems [11]. We believe so that he will be very useful to incorporate personalization for composing Web services.

Section 2 presents the related work describing the personalization approaches for web service composition. Section 3 presents an overview of the proposed approach, the user profile orientation, the used ontologies, and the personalization based knowledge. Section 4 explains the user's profile construction. Section 5 presents the user's request personalization process. Section 6 treats the personalization of composition services. After that, we present an illustrative example. And we finalize by experimentation and evaluation of the proposal.

2. Related work

Many existing approaches in the literature treat the concept of personalization for Web service composition. In this regard, (Fekih et al) [4] [present an approach that is both semi-automatic and semantic one hand, user intervention is necessary, where it is represented from its profile and its preferences. Furthermore, the service selection is made up of a semantic description based on OWL-S [12]. It is also important to note that the authors present the service selection process in three stages namely, the query expression that integrates the user profile where this last based on real information (name, date of birth). The service discovery. The final step is the validation of the research by the user that declares whether the user is satisfied or not, knowing that failure to user satisfaction, the whole process will be repeated.

Shanchen et al [5] thinks that a context classification is important. We distinguish then between the U-context (user context), the W-context (web service context), and R-context (context Resource). On the one hand, the context classification allows to better establishing customization, and consistency can check the status of a Web service after being personalized. On the other hand, the three classes classified are interconnected. The user is the so most dynamic component: its needs, its preferences, and its conditions always vary. The resource is the stable component which characteristics and constraints can be known in advance.

Mcheick H et al. [13, 14] present a crucial adaptation of Web services face the changes that may affect them. The approach aims to resolve Web service adaptation problem. It is based on the principle of adding two components named "Manager of appearance" and "context manager".

Not to forget, to include correspondence from ontologies based on lexical database (Word net) [15]. In [6], authors believe it is useful to store the context before its release in the selection process and the composition of Web services. This therefore provides a rich and reliable representation of data captured in the form of ontology, based thereafter on mathematical formulas for algorithms of semantic correspondence [16].

Given that a composition of services intended to meet the need for a user. Based on this principle, and thus returning the compositional approaches presented in the previous section, we note that regardless of the approach proposed, it still lacks personalization throughout the composition process from the user's request pending the outcome of composite service. Indeed, we choose the user profile as a medium to introduce personalization.

The following section provides an overview of the proposed approach by presenting the choice of user's profile, the choice of the used ontologies, and the choice of personalization's forms.

3. A general overview

This section presents first an overview of the proposed approach. Second, it explains the choice of user profile. Third, it presents the chosen types of ontologies. And finally, we present the different chosen forms of personalization.

This architecture provides a global perspective that concerns personalization of the request, and the development, management and operation of Web services. Thus, it essentially consists of three layers. As shown in **Figure 1**, the first one presents a user profile construction process based on ontologies. The second layer is the personalization of the user's request, which consists of the evaluation of the request based on the constructed user profile. The last layer consists in the personalization of composition processes using also the user profile.

3.1 Orientation user profile

Currently, user profiles play a very important role in all digital environments [17]. Profile integration is one of the ways systems that can be adapted to users in digital environments. Each information system based on services or services composition should primarily support resources in order to meet the changes in system use required.

The user model is a representation of information about an individual user that is essential for an adaptive system to provide the adaptation effect. From there, we define a user profile as "*necessary information may be necessary to guide the user's request personalization*". So, personalization is defined by a set of specific individual

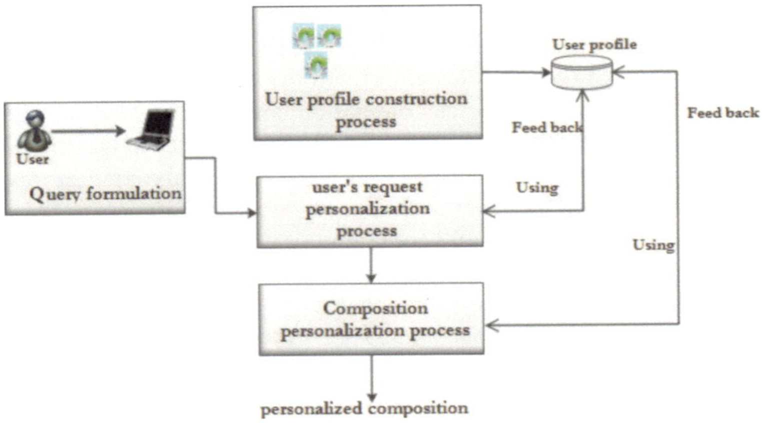

Figure 1.
Overview of the approach.

preferences for each user. As the profile includes data collected from users that are effective in evaluating a request, then we relied on the profile technical based on ontologies; we choose to use a short-term profile constructed from the domain ontology and a long-term profile constructed from the ontology of the user. We will explain more in what follows the process of user's profile creation.

3.2 The used ontologies

When we talk about data representation, we are looking for tools that can enrich and strengthen the efforts of these. Thus, we work in a dynamic environment; we need a reusability of knowledge. Besides, for our approach, we need a tool for indexing and information retrieval. This is why our choice is for the ontology.

Our approach is based on: domain ontology and user ontology. The domain ontology is an indispensable resource for the personalization process. Indeed, the more the context of the application, the domain is also an important factor for a personalization process. For the ontology of the user, since the choice of data is very important insofar as it specifies the information needed to present a user and its preferences, we are based on the multidimensional approach (Kostadinov D) [10] in the first place, then we have taken the approach of (Katifori) [18] which allows the passage of data about the user to the concepts representatives in the user's ontology thereafter.

3.3 Personalization based on knowledge

Personalization can take many forms; include in this the "result of filtering" often known by eliminating unwanted data rather than looking for specific data within the same document flow. A second form is "the query's enriching" where personalization is defined as learning achieved from the preferences given by the users for the reformulation of the query thereafter.

Seeing that we need a reformulation of query and search for specific data, we retain the "the request enrichment". Personalization is then defined as learning made by users for the reformulation of the request thereafter. Indeed, in our approach, the query undergoes enrichment from the two ontologies (domain ontology and user's ontology).

4. User's profile conception

For the implementation of a user's profile technique, we essentially based on two approaches: **(1)** The multidimensional approach (Kostadinov D) [10] and **(2)** (Gauch S) approach [19]. This choice is explained by the fact that these two approaches have two complementary forms of personalization. Thus, (Kostadinov D) approach [10] aims to propose a set of open dimensions, able to accommodate most information characterizing Profile. It is based primarily on seven dimensions that are (personal data, center of interest, domain ontology, the expected quality of the delivered results, personalization, security and confidentiality) (**Figure 2**).

Otherwise, since the classification, organization and structuring of profile data is a key element of personalization, so we have chosen Gauch [19] approach which aims to create a profile of ontology-based user without using the user interaction based on a classification of concepts.

To set up a personalization process, it is essential to choose its type to apply. Two main issues arise in this respect; the first issue deals with the dimensions of user context and the second addresses the choice of personalization's form. We are interested in the first question, so it is useful to study a user's context before proceeding to the description of the dimensions for the latter.

Since a context is composed of several dimensions [20], we distinguish a *social dimension* that describes the potential membership of the user. It may be; individual, group or community. A time dimension that is interested in the temporal context of the need, we thus distinguish between a short-term intention and a long-term intention. The first type is related to the needs and preferences of the user during a search session for information, while the long-term context (personalization) reflects the needs and persistent user preferences [21]. Finally, an application dimension is interested in application area.

Regarding the dimensions, we have the following choices: Personal data, this dimension is composed of two parts: a static part of the profile and a dynamic part. The static part concerns the following three: (1) " the user's identity", (2) " demographic data", (3) "physical description". The dynamic part subclass concerns the sub-class "category". For the center of interest, a user may be interested in several concepts. Indeed, we do in this user's modeling framework differentiation between its various needs. We will explain in the next section the used ontologies (**Figure 3**).

In order to highlight the user profile, we illustrate this using the example shown below. Thus, a profile can be defined by the following concepts: Identity which defined by the first-name (Philippe), and the last-name (Arno). The second concept is U-profession specifies the user's work (nurse). U-Civil-status indicates if the

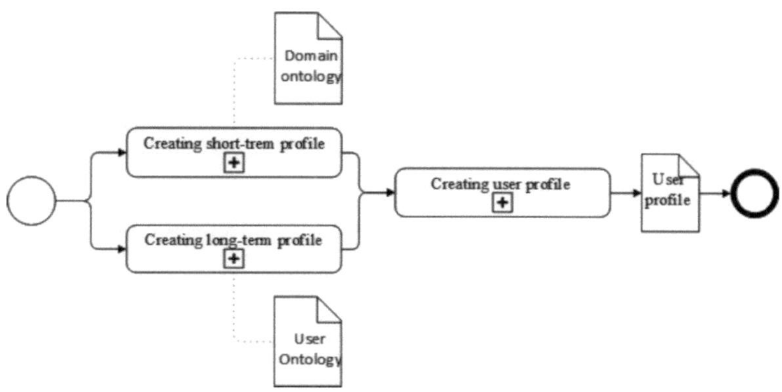

Figure 2.
Process of user's profile conception.

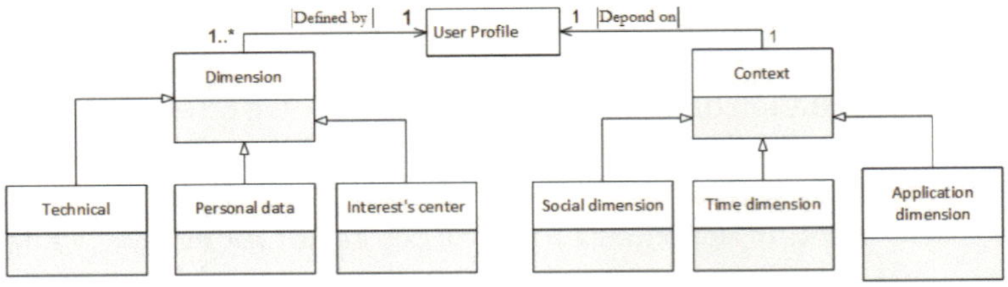

Figure 3.
User profile model.

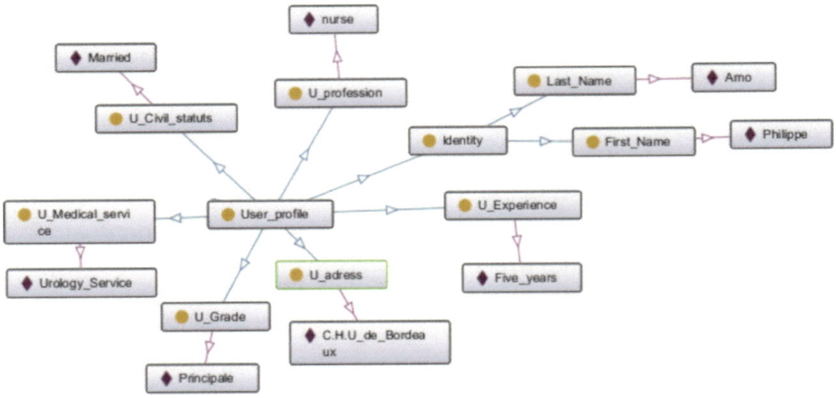

Figure 4.
User profile example.

user is married or not. U-experience specifies the period of work as a nurse (Five years). The concept U-Medical-service specifies that the nurse works in the urology department. U-address presents the nurse's workplace (C.H.U-de-Bordeaux). Finally, U-Grade specifies what nurse's type is involved, which is linked to the experience (**Figure 4**).

5. Personalization of web services composition

The personalization of Web services composition process starts from "User's request personalization" process which based on the constructed user's profile. It allows the user to express their needs, by performing first, a (first enrichment) query from a short-term user profile. After the deployment of data acquired, the application undergoes a (second enrichment) from the data of a long-term profile leading subsequently to a personalized query, which will be the basis of the next layer. In this way, an end-user profile based on ontologies is constructed. Finally, we should mention that if a user needs to make updates to some data, it uses the user profile.

The services composition processes is essentially based on 3 steps which are: decomposition into subqueries, discovery and selection of services and proposal of composition's plans. For our approach we followed the same steps by adding the notion of personalization and we choose a semantic Web services. Thus, the dynamic process proceeds as follows. The process starts from a personalized query to invoke as the end the services necessary to meet the expressed needs. It gives us output, a set ordered services for execution. Thus, the process started from a decomposition and verification of sub-queries. In fact, this is based on a comparison of the profile parameter which had been added to the request and the WS Context parameter. The second step is a personalized discovery and selection for Web services: This phase is based on a semantic description for the discovery and relevant selection of personalized Web services (**Figure 5**).

Given that a personalized query is identified by three parameters (InpReq, OutRep, ProfReq) and a Web service (WS) is identified by (Input, Output, Context), so actually the selection of services in a personalized way is as follows:

- First, check the compatibility of inputs of a personalized query and inputs of a Web services.

- Secondly, check the compatibility of the outputs of a personalized query and outputs of a Web services.

- Thirdly, check the compatibility of the profile for a personalized query and context of a web service.

If the comparison between the different parameters is validated, so the process goes to the next step which is "the personalized services composition".

From this validation is calculated the similarity between the different settings using the algorithms presented below. Thus, the services were selected in a non-predefined order, so they were selected and ordained dynamically to meet the needs of the user.

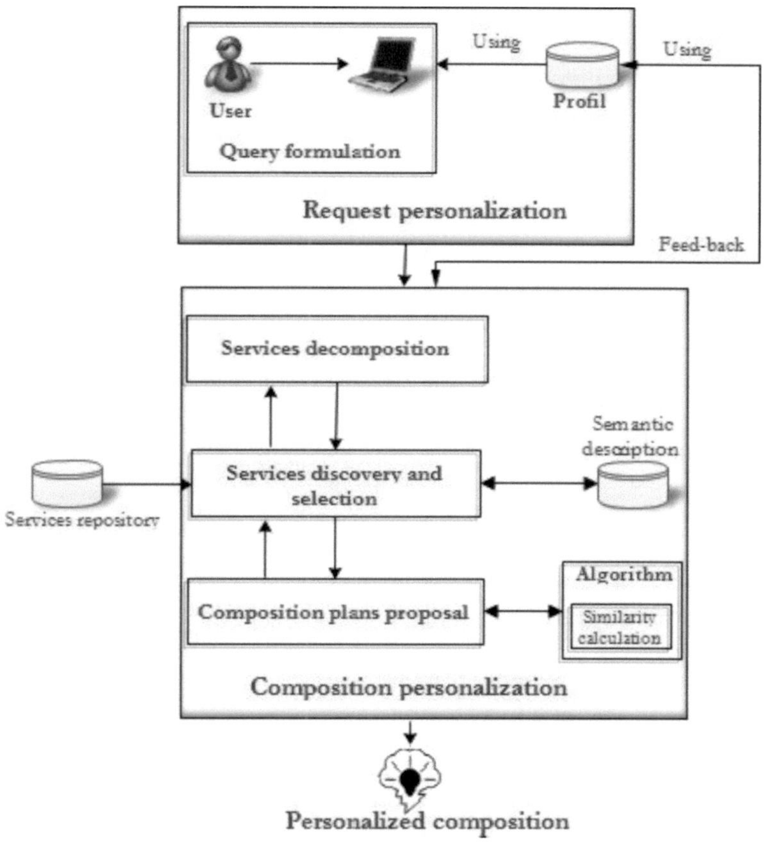

Figure 5.
Architecture of the proposed approach.

If the comparison between the three query parameters and WS is not validated so we will go only to compare the inputs, and the profile with context in the same way mentioned. If that is validated, the result is satisfying, if not; we will compare the profile with the context. If it is approved, so we have a proposed composition plans, referring to ontology "Word net", if not there's no result.

As a consequence, the new parameters (Profile and Context) that we have added are responsible for a personalized Web services composition or not. In this way, the process makes a composition of relevant personalized services in choice, in number and scheduling services. To clarify the proposed approach, we present on the following section an illustrative example of medicines circuit.

Algorithm 1: Discovery module

Inputs: SRq, bestsw
Outputs: SWF
Taux-sim ← 0
For each s in SW **do**
If similarity (SRq, s) ← Taux-sim ***so***
 Taux-sim ← similarity (SRq, s)
  SWF ← s
End if
End for
 bestsw← Better(SWF)
returns SWF

Algorithm 2: Semantic similarity.

Inputs: SW, SRq
Outputs: Taux-sim.
Taux-sim ← 0
If (EntReq=EntSW) *and* (SortReq=SortSW) and(ProfReq=ContxtSW) ***so***
 Taux-sim ← 3
else if (EntReq) included (EntSW) and (SortSW) inclus (SortReq) and (ProfReq) included (ContxtSW) ***so***
 Taux-sim ← 2
 Else *if* (SortReq) included (SortSW) and (ProfReq) inclded (ContxtSW) ***so***
 Taux-sim ← 1
 else
 Taux-sim ← 0
End if

---

6. Illustrative example

To highlight the interest of the proposed approach, we present a case study of a medicines circuit application in a health facility.

Healthcare organizations are highly dynamic working environments which are facing the challenge of delivery personalized services to their patients in a very cost-effective and efficient way. Many reports in the healthcare field state that there is an "absence of real progress towards applying advances in information technology to improve administrative and clinical processes" [22]. Furthermore, in healthcare organizations, the lack of personalization of contemporary enterprise information systems is considered as a major obstacle in improvement of organizational and medical treatment processes.

Let us start with the following case in which the clinician writes a nominative prescription. The pharmacist valids the prescription after a pharmaceutical analysis. Drug doses issued to the clinical are unitary and nominative (ready to be administered to the patient) and respect the administration plan for the next 24 hours. The renewal of the grant by 24-hour period is provided by the pharmacy (chemistry) if the prescription is still valid.

This scenario presents a situation where the actor tries to satisfy a need for a complex query. There is an executable composition to satisfy him. Indeed, as part of his mission, he must prepare medicines to administer for its care unit patients from a stock.

This activity is usual for the nurse/pharmacist (chemist), to know arrangement of patient medicine and the preparation according to the medicines type.

We suppose that we have a certain number of services as shown in **Table 1**, in a repository in which we distinguish.

In the following, we present the process of medicines preparation in the case of a personalized composition and a non-personalized one.

6.1 Query expression step

To specify his purpose, the actor can formulate his need as follows by expressing his purpose. Thus, given the following request, the preparation can therefore be carried out as follows. Preparation by medicines type and arrangement per patient.

User's query: Prepare(medicines-list, doses-list, date, care-unit-list) ∧ Arrangement(care-unit-list, patient-list).

We distinguish between two different requests; a non-personalized request and personalized one, knowing that the personalized request is enriched by a user profile which contains the following information (profession, experience) (**Table 2**).

Thus, in our approach we defend that the personalization must start since the expression of the request because it allows giving more information about the user and his environment. This will make it possible to define a more specific context for each actor.

Service's number	Service's name	Service's role
S1	Pharma preparation	Medicines preparation according to their types
S2	Pharma arrangement	Medicines arrangement for each patient
S3	Pharma edition	Save and print package
S4	Pharma verification	Verify authorized doses for each patient
S5	Pharma attribution	Assign medicines for each patient
S6	Pharma compatibility-verification	For each patient, check the compatibility of medicines

Table 1.
Selected web services.

Non-personalized request	Personalized request
Prepare([Brilinta/180 mg, Amiodarone/120 mg, Digoxine/160 mg, Plavix/300 mg, Vérapamil/250 mg], doses-list, 26/11/2016, urology) ∧ Arrangement(urology, [Paul/Brilinta/180 mg, Frederic/Amiodarone/120 mg/, Paoula/Digoxine/160 mg, Celine/Plavix/300 mg, Maria/Vérapamil/250 mg])	Prepare([Brilinta/180 mg, Amiodarone/120 mg, Digoxine/160 mg, 1111Plavix/300 mg, Vérapamil/250 mg], doses-list, 26/11/2016, urology) ∧ Arrangement(urology, [Paul/Brilinta/180 mg, Frederic/Amiodarone/120 mg/, Paoula/Digoxine/160 mg, Celine/Plavix/300 mg, Maria/Vérapamil/250 mg]) ∧ ∃ user(nurse, [5 years = a block nurse, 6 years = referring nurse])

Table 2.
Presentation of the two request types.

6.2 Research and services selection step

Having doubts about the choice of services. The user resorts to the automatic selection based on a repository of services and a basic knowledge (which is an example of a composition plan). Thus, for the services rendered by a non-personalized request, we will obtain the services S1, S2, S3, S4, S5, and S6. This is explained by the fact that the profession's actor is not defined. Therefore, the system offers all services. However, for the services rendered by a personalized request, based on the profile parameters and as a nurse she has the right to perform services S1, S2, S3, and S6. In addition, for "Pharma verification-compatibility" service; the pharmacist has no right to verify medicines compatibility for a patient. But rather, checks the compatibility between medicines (**Table 3**).

6.3 Services composition step

Services composition result is based on a similarity measurement algorithm and a dynamic discovery between different parameters. Starting from a non-personalized request (**Figure 6**), the system proposes the following composition plans. A nurse must first verify the authorized doses for each patient. Then, she prepares medicines by their types. After that, she attributes for each patient his mentioned medicines. The next service allows arranging medicines according to each patient. On Urology-unit and through "PHARMA Compatibility-Verification", the system allows knowing the compatibility between Paola and Digoxine. Finally, the actor saves the order.

For personalized query (**Figure** 7), through profile integration, the system proposes three services for a nurse. Thus, from her profession and experience, she can

	S1	S2	S3	S4	S5	S6
Results of non-personalized request	✓	✓	✓	✓	✓	✓
Results of personalized request	✓	✓	✓			✓

Table 3.
Service research results for both types of queries.

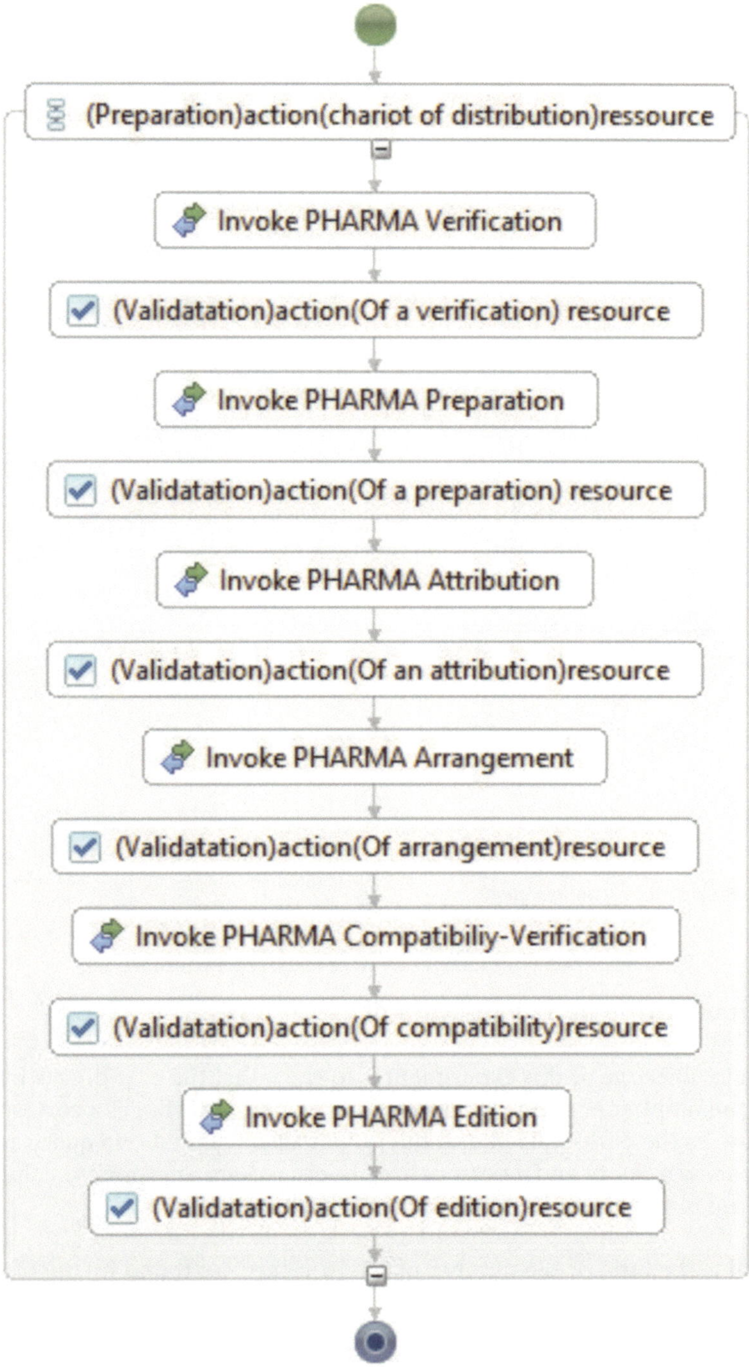

Figure 6.
Composition result for a non-personalized query.

begin with the preparation of medicines by their types. Since this actor has already worked as a referring nurse, the system proposes for her to arrange medicines for each patient. Finally, the last service is "PHARMA Edition" which allows saving the order.

In that way, we have proved that the profile integration influences to have a personalized request that results in an impact on research and selection of services in order to obtain a personalized composition.

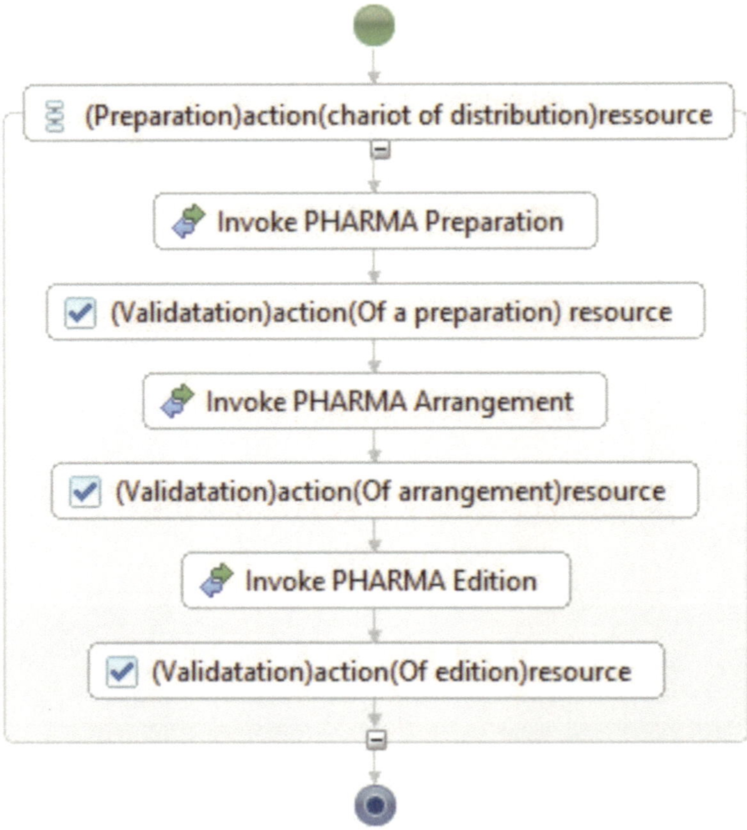

Figure 7.
Composition result for a personalized query.

7. Experimentation and evaluation of the proposal

The main objective of this experiment is to show that the user profile integration increases the number of relevant services. Thereafter, we will get a corresponding composition to the personalized user query. For this, we used two query types; a non-personalized query and a personalized query, where appropriate valuation measures are mainly based on precision and recall.

- **Recall** is the ratio of the number of relevant services found by the filter to the number of relevant services available.

Recall = Number of selected relevant services / (number of relevant selected services+ number of relevant non-selected services).

Based on **Figure 8**, we notice that the recall for a personalized query is high compared to a non-personalized query. This is due to the recall which is always the number of relevant selected services over all relevant services. So the user will have access to information that he wished to have.

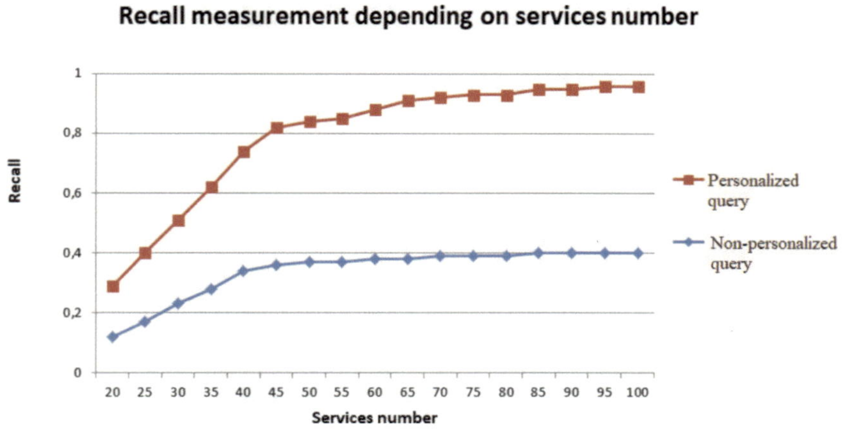

Figure 8.
Recall measurement curve.

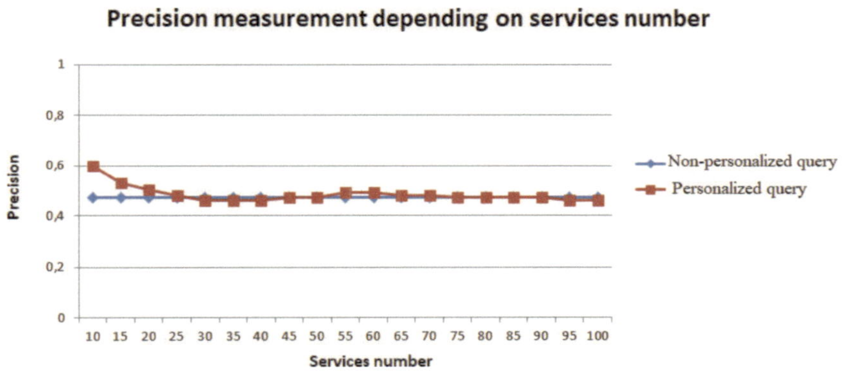

Figure 9.
Precision measurement curve.

- **Precision** is the proportion of relevant services among the selected services.

Precision = Number of relevant selected services / (number of relevant selected services + number of non-relevant selected services).

As shown in **Figure 9**, the accuracy for a personalized query is high compared to non-personalized. However from a number of services equal to 27, the curve for a personalized query is the same for a non-personalized query. This is explained by the precision which is the number of relevant selected services over all services.

8. Conclusion and perspectives

This paper presents a dynamic approach for personalization of Web services composition. First, the construction of a user profile from domain ontology and user's ontology is a key point for personalization, which leads to the construction of a personalized query where each user may have personal data that is stored in a parameter named "Profile". On the one hand these data facilitate the personalization subsequently, hence the construction of a corresponding request to the need of the user. On the other hand, the process is based on a similarity measure algorithm for the personalized discovery of web services, which allows thereafter establishing a personalized composition.

This approach has a dynamic modeling for user not only for the query expression but also the composition process.

We should note that a scaling test is in progress as Web services adapted to our needs are not available. So, we were forced to edit them manually, which took a long time. But we are working on the scaling (we increase the number of services and we observe the result).

As a future work, we can identify two interesting perspectives. The first one is How to improve the results relevance in terms of selected services. The second one is how to respond to business needs in dynamic contexts.

The second perspective is to design compositions that integrate a personalization for business process satisfaction and user satisfaction.

Author details

Sarra Abidi[1]*, Fathia Bettaher[2] and Myriam Fakhri[3]

1 ESPRIT, Tunisie

2 IRISCOM Hatem Betaher laboratory, ISIMG, Tunisie

3 LSIS Laboratory, Aix Provence University

*Address all correspondence to: sarralabidi@gmail.com

References

[1] Abidi, S., Essafi, M., Fakhri, M and Ben Ghezala H. "Refined cloud services discovery based on user model including privacy-preserving access control", Knowledge Based and Intelligent Information and Engineering Systems, (2018).

[2] Amina, A., Mohammed, B. and Driss Chenouni, "A phased two stages conceptual framework for web services composition", Journal of Engineering and Applied Sciences, Vol. 13 No. 7. (2018)

[3] Fekih, H., Mtibaa, S. and Bouamama, S. "User-centric web services composition approach based on swarm intelligence", IEEE 18th International Conference on High Performance Computing and Communications. (2016)

[4] Fekih H, Mtibaa S, Bouamama S. «An Efficient User-Centric Web Service Composition Based on Harmony Particle Swarm Optimization». International Journal of Web Services Research · January Volume 16 . Issue 1 . (2019)

[5] Shanchen P, Qian G, Ting L.H, Guanguan X, Kaitali L. « A Behavior Based Trustworthy Service Composition Discovery Approach in Cloud Environment ». IEEE Access (Volume 7), (2019)

[6] Yuan Yuan, Weishi Zhang, Xiuguo Zhang. « A Context-aware Self-adaptation Approach for Web Service Composition ». 3rd International Conference on Information Systems Engineering. (2018)

[7] C Peltz.. Web services orchestration and choreography. IEEE Computer, pages 46-52.(2003)

[8] A Dumas. M Oaks. P Barros. Standards for web service choreography and orchestration: Status and perspectives, Procs of the 3rd International Conference the Business Process Management (BPM 2005), 1st International Workshop on Web Service Choreography and Orchestration for Business Process Management, Nancy, France , pages 1-15. (2005)

[9] S Dustdar and W Schreiner. A survey on web services composition. Web and Grid Services, Vol. 1, No. 1, pages 1-30. (2005)

[10] Kostadinov D. Personalization of information a profile management approach and reformulation of queries. Thesis in Computer Science. University of Versailles Saint-Quentin-En-Yvelines. (2005)

[11] Serge Garlatti and Yannick Pri, Adaptation and personalization in the Semantic Web, Revue I3 Information-Interaction - Intelligence, 24 pp. (2004)

[12] Fekih, H., Mtibaa, S. and Bouamama, S. (2016), "User-centric web services composition approach based on swarm intelligence", IEEE 18th International Conference on High Performance Computing and Communications.(2016)

[13] Mahmoud Zakaria Maamar. Qusay H. Soraya Kouadri Mostéfaoui. Context for personalized web services, Proceedings of the 38th Hawaii International Conference on System Science. (2005)

[14] Amel Hannech. Hamid Mcheick. Semantic web services adaptation and composition method,Department of Computer Science and Mathematics University of Quebec at Chicou- timi (UQAC) Chicoutimi, Canada (ICIW: The Eighth International Conference on Internet and Web Applications and Services). (2013)

[15] George A. Miller, WordNet: a lexical database for English, Magazine

Communications of the ACM CACM Homepage archive Volume 38, pages 39-41. (1995)

[16] Amel Hannech, Hamid Mcheick, Composition model of semantic web services orchestration on demand, Department of Computer Science and Mathematics University of Quebec at Chicoutimi (UQAC) Chicoutimi, Canada. (25th IEEE Canadian Conference on Electrical and Computer Engineering (CCECE)). (2012)

[17] Brusilovsky, P., Millàn, E, "User models for adaptive hypermedia and adaptive educational systems", "The Adaptive Web", page 53, (2007)

[18] Golemati M. Halatsis C. Katifori A. Lepouras G. Vassilakis C. creating an ontology for the user profil: Method and applications. Proceedings of the First IEEE International Conference on Research Challenges in Information Science (RCIS). (2007)

[19] Challam V. Chandramouli A. Gauch S. contextual search using ontology-based user profiles, Proceedings of RIAO, Pittsburgh USA. (2007)

[20] Bouzeghoub M, Kostadinov D, Personalization of information: Overview of the state of the art and a flexible model profiles definition, Conference in Search of Information and Applications (CORIA), Grenoble, France. (2005)

[21] Fuhr N. information retrievial: introduction and survey, postgraduate course on information retrievial, University of Duisburg-Essen. (2000)

[22] Lenz, R and Reichert, M, "IT support for healthcare processes – Premises, challenges, perspectives," pp. 39–58, (2007)

Section 2

Artificial Intelligence and Affective Computing

Chapter 3

Humanistic Next-Generation Artificial Intelligence Capable of Association

Seiji Tsuchiya

Abstract

The third artificial intelligence (AI) boom focused on the "handling of large amounts of data" and "automated learning." One may think that AI can do anything because it is capable of automated learning, but there are still many problems that AI must tackle. The "necessity of a large amount of data" is and will become an even more significant problem. Obtaining an accurate solution from small amounts of data requires imagination and the detection of trends from a small number of phenomena. One approach is to add artificial data. For example, data can be created by intentionally including noise, and the variation may be expanded by a crossover. Different data can be generated by association or inference. Needless to say, these are artificial data and are not correct cases. "Humanistic AI" must be implemented by devising a scheme to allow accurate learning from small amounts of data. I think that the days when robots are considered enemies are transient and robots will soon be recognized as good partners that support humans instead of being rivals.

Keywords: artificial intelligence, humanistic AI, association, inference

1. Introduction

The third artificial intelligence (AI) boom focused on the "handling of large amounts of data" and "automated learning." Collecting and learning information expressed in various forms in a variety of fields can lead to the automated detection of knowledge and rules. One may think that AI can do anything because it is capable of automated learning, but there are still many problems that AI must tackle. The "necessity of a large amount of data" is and will become an even more significant problem.

Large amounts of data are indispensable to statistic processing. In other words, problems where data cannot be collected cannot be solved. One view is that "problems where the collection of data is difficult are not big problems anyway, and therefore, can be ignored." Is this really acceptable? Ignoring a problem because the amount of data is small separates problems into those that can and those that cannot be processed using AI. Stretching in this direction leads to the formation of gaps, which then leads to discrimination and information gaps. Therefore, although an effective use of information technology should be accessible to the weak, a totally opposite scenario may occur.

2. Problems in Japan

The population of Japan is approximately 120 million, which is about one tenth of that of all of the Western countries combined or that of China. As a consequence, the amount of data that a large country can collect in 1 month takes around a year to collect in Japan. Therefore, Japan cannot win against large countries, as the performance of AI is currently governed by the amount of collected data.

Therefore, AI developed abroad will be imported and used. This is sufficient to some extent from a macroscopic point of view. However, Japan is a very unique group as compared to large countries, a fact that is often overlooked because Japan is highly ranked, for instance, in terms of economy. Many aspects of the Japanese culture are peculiar; hence, there will be some unreasonableness if AI developed abroad is used as-is. From a microscopic point of view, Japanese people have to put up with some issues in their lives.

For example, a translation service that a large company provides for free mistranslated “Sakaisuji Line,” which is a subway line in Osaka, as “Muscular Line,” and “Sakaisuji” as “thigh muscle.” This is a big problem for people living in Japan, and international sightseeing visitors will be bewildered (**Figure 1**) [1].

This type of mistake can be easily detected and resolved by Japanese people. In contrast, for companies providing services globally, a small mistake in a specific region in a single country might not be an issue. Indeed, this mistranslation must not have been viewed as a problem that needed to be addressed, because no correction was made for a long time after this problem was reported. Thinking from a global macroscopic viewpoint naturally emphasizes covering more countries, regions, and languages. However, minority groups must clearly understand this issue: systems are not always created and provided with sufficient consideration to minority groups.

Figure 1.
Example of a translation service.

3. Limits of statistical processing

Having more data is not always better. For example, consider someone who loves “tomatoes” very much and eats tomatoes every day. A favorite friend’s birthday is approaching, and he wants to give this friend a present. After consultation with various acquaintances, the recommendation was to “give my favorite thing as a present.” He therefore decided to give “tomatoes” as the birthday present because “I love ‘tomatoes’ so much; I eat them every day.” Is this a good idea?

Of course not. Giving "tomatoes" as a birthday present clearly contradicts common sense. However, "giving my favorite thing as a present to others" is not a wrong idea. Had the person loved "cakes" instead of "tomatoes," then giving "cakes" as a birthday present to his favorite friend would have been a very successful result. Then, what is the difference between "tomatoes" and "cakes"? These are both food! Distinguishing these just by using statistical processing is very difficult. In contrast, the "tomatoes" lover might be very happy when very rare, top-grade tomatoes are given to him as a birthday present.

As is evident from this example, schemes that do not simply analyze large amounts of data but instead analyze high-quality data or give solutions with good precision will be indispensable.

Many data can be collected if the target is large. One resolution when a large amount of data cannot be collected is to increase the denominator by increasing the target domain. A considerable amount of data can be collected in this case, but a large amount of irrelevant data will also be collected. In contrast, purging unnecessary data that become the noise leads to the narrowing of the target domain. Characteristic data can be collected by focusing on a certain domain, but this results in the problem of how to collect large amounts of data in the limited domain.

The issue of quantity and quality is basically a trade-off relationship. Collecting large amounts of data is necessary to cover various cases and to judge the importance of certain things by investigating the frequency. Therefore, the shortcut to obtaining high-quality data is to secure the variation and sufficient data for determining the relative importance.

4. Mechanism to derive the solution from less data

Obtaining an accurate solution from small amounts of data requires imagination and the detection of trends from a small number of phenomena. One approach is to add artificial data. For example, data can be created by intentionally including noise, and the variation may be expanded by a crossover. Different data can be generated by association or inference. Needless to say, these are artificial data and are not correct cases. However, if an environment where very delicate simulations are possible can be obtained, artificial data that are not true, but very close, may be created.

We humans can, without actually experiencing everything, imagine and think by reading books or listening to others' experiences. Simulated experience is valuable. I think that computers can also respond to simulated data. However, computers are not as imaginative as humans. Common sense is necessary when imagining things. The timing, circumstances, and situation must be taken into account. Moreover, judgments must consider the position of the counterpart, human relations, the atmosphere, and the underlying background.

However, current computers are not capable of this task, which is understandable because computers live in a world of "0"s and "1"s and things are considered and handled as symbols. AI that can support humans and can be active as partners need a mechanism that has common sense and can understand and share human feelings.

5. Implementation of humanistic AI

I founded the Artificial Intelligence Engineering Research Center at Doshisha University in 2018, and I am now its Director. Many professors who focus on various AI-related research work at this research center and are conducting research on AI from diverse viewpoints. I am particularly interested in the "implementation of humanistic AI" [2].

"Humanistic" has two aspects: one is "common knowledge," which is "what everyone knows," and the other is "common sense," which is "conscience and sound consideration and judgment." "What everyone knows" can be statistically processed because it can be found explicitly in dictionaries or is what many people agree upon. However, "conscience and sound consideration and judgment" is very tricky. It depends on ethics, morals, manners, virtues, and cultures; thus, the "correct" answer is vague, and judgments can vary from person to person. However, there can be some guidelines such as "this is not good" or "this is impossible" (**Table 1**).

Examples of expressions that the Japanese use casually are "the size of 20 Koshien Stadiums" and "input single-byte alphanumeric characters in this field." These expressions are not strange to the Japanese but are difficult to understand for people outside of Japan. A person unfamiliar with the size of the "Koshien Stadium" cannot relate to what the term "the size of 20 Koshien Stadiums" means. "Single-byte alphanumeric characters" are different from "double-byte" characters, and this expression does not make sense to Westerners who do not use double-byte characters that appear in languages such as Japanese and Chinese. Another example is the following conversation of a married couple: "Help with housework when you are off from work!" "I'm ceaselessly driving a truck, so please let me rest on those rare days off!" The assumption that "the wife is doing the housework, and the truck driver is the husband" is a very outdated common sense in the current world of gender equality.

"Conscience and sound consideration and judgment" therefore changes with, for example, the timing, circumstances, sex, age group, region, position, and/or era. Trying to learn this automatically results in a lack of data. The population of Japan is already small, and classifying data by sex, age group, or region further reduces the amount of data. However, "humanistic AI" must be implemented under these circumstances by devising a scheme to allow accurate learning from small amounts of data (**Figure 2**).

	Common knowledge	**Common sense**
Meaning	What everyone knows	Conscience and sound Consideration and judgment
Definition	Found explicitly in dictionaries	The "correct" answer is vague
Judgments of people	Many people agree upon	Judgments can vary from person to person

Table 1.
"Humanistic" has two aspects.

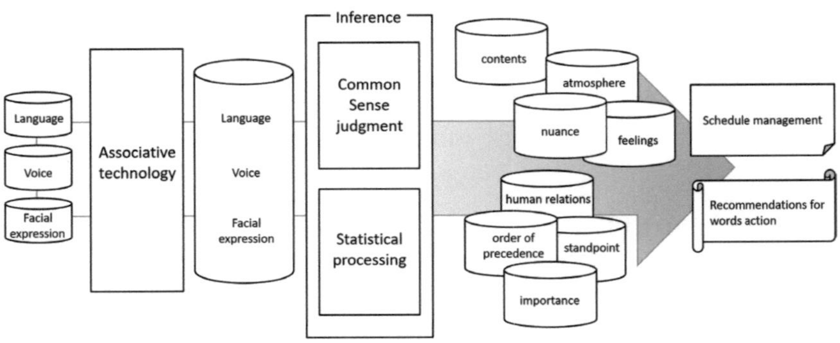

Figure 2.
Example of "humanistic AI."

6. Estimation of emotions from language

My specialty is AI with an emphasis on natural language processing. In particular, I have continued to study emotions ever since I started research. With the widespread use of smartphones, speech input is gaining popularity again. New search methods by the use of Siri in iPhones and smart speakers have been proposed, and these are becoming accepted. Further advances in technology and the day when everyone can freely use sophisticated computers are being anticipated. One solution is to use the means of communication between humans as the means to use computers. When robots that can coexist and live together with humans are developed, the ability of robots to understand humans, judge human emotions, and sympathize with humans should be a very important element. Therefore, I focused on reading the emotions of the counterpart from the contents of one's speech and proposed a method to judge the speaker's emotions by analyzing the contents of the speech (**Figure 3**) [3–5].

Emotions were judged on the basis of a knowledge database where emotions were defined for 406 combinations of 203 objective word categories and two verb and normal and nominal adjective word categories, or for 8024 combinations of 34 objective word categories; 59 verb and normal and nominal adjective word categories; active or passive voice (two categories); and positive or negative form (two categories). When categorizing objective words and verb and normal and nominal adjective words, the ambiguity is judged from the relationships between words (**Figure 4**). To effectively use this knowledge database where a small amount of knowledge is registered to judge processing and emotions, I developed a proprietary "concept base" that automatically interprets limited knowledge broadly to respond to diverse expressions [6–9] (**Figure 5**). One result showed that this proposed method is capable of reproducing 74.2% of the emotion judgment ability.

The "concept base" has been developed for over 20 years by my research group. "Word2vec," which has gained considerable popularity recently, is a dictionary built on a similar concept [10]. Word2vec appeared recently; thus, my feeling is that "the times have finally caught up with our ideas." Word2vec is a method that expresses the meaning of a word using multiple numbers on the basis of the hypothesis that "the meaning of a word can be characterized using words that collocate with

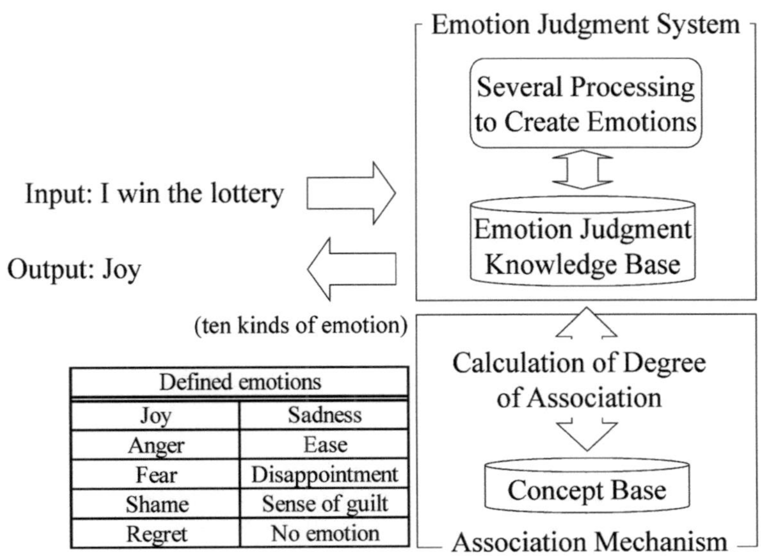

Defined emotions	
Joy	Sadness
Anger	Ease
Fear	Disappointment
Shame	Sense of guilt
Regret	No emotion

Figure 3.
Outline of the emotion judgment system.

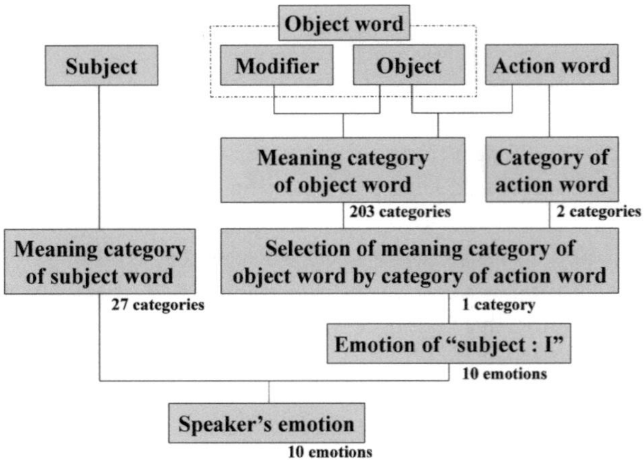

Figure 4.
Flow of emotion judgment system.

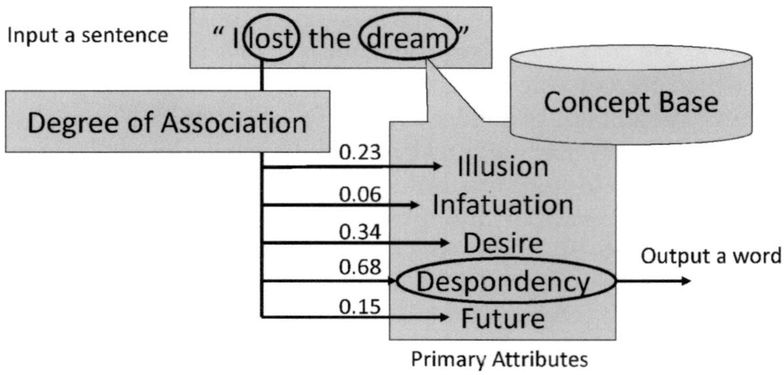

Figure 5.
Technique of automatically interpreting limited knowledge broadly to respond to diverse expressions.

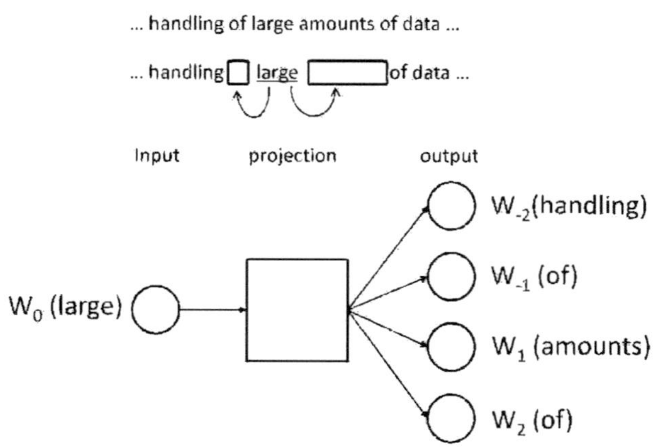

Figure 6.
Image of Skipgram in Word2vec.

it." In other words, a word is expressed using a vector in an arbitrary dimension. Expressing the meaning with multiple numbers requires learning from large amounts of data. One method is "Skipgram" that learns words close to a given word (**Figure 6**), and another is "CBoW," which learns words that often appear when a certain number of words exist (**Figure** 7).

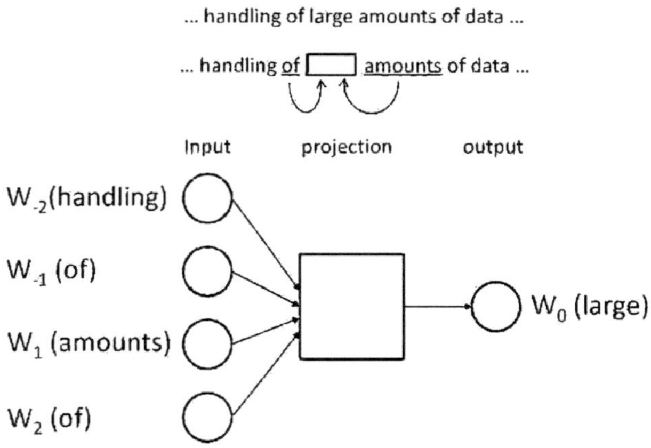

Figure 7.
Image of CBoW in Word2vec.

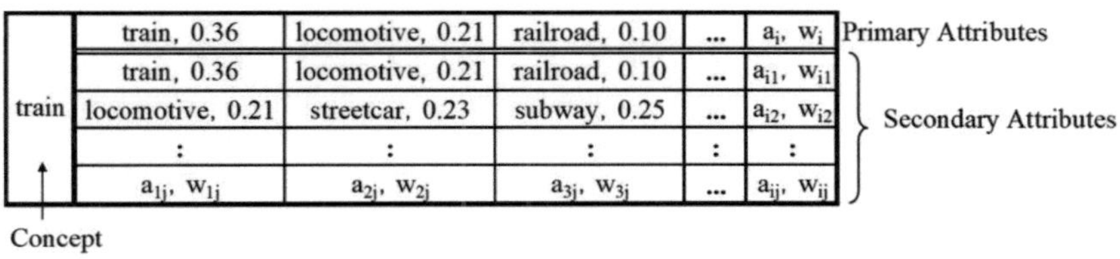

Figure 8.
Example of concept base.

	Concept base	Word2vec
Accuracy	89.9%	78.4%

Table 2.
Result of accuracy comparison between concept base and Word2vec.

Word2vec needs to arbitrarily specify a finite number of dimensions. Therefore, the meaning of words must be compressed, and the expressions are slightly forced. In contrast, the concept base expresses the meaning of words with multiple words, which is different from Word2vec. This setup allows the expression of a word with multiple, or, in theory, an infinite number of words. Therefore, all the concepts of a word pictured in a human brain can be expressed (**Figure 8**). Indeed, our concept base captures the meaning of words more precisely than that captured in Word2vec (**Table 2**) [11].

7. Conclusions

The implementation of "humanistic AI" may lead to humans falling in love with a robot instead of a real human. A robot may know more about you than other humans, will do what you want to be done, will not complain, and can have a good conversation. Therefore, the chances of falling in love with a robot are considerably high.

I think that the days when robots are considered enemies are transient and robots will soon be recognized as good partners that support humans instead of being rivals. When such a time comes, what will humans do and what should humans do? Research on AI forces me to think about humans instead of just on technology. After all, the role model of a robot is none other than us humans. I wish for an evolution to a world where humans can be humans.

Author details

Seiji Tsuchiya
Doshisha University, Japan

*Address all correspondence to: stsuchiy@mail.doshisha.ac.jp

References

[1] https://www.itmedia.co.jp/news/articles/1903/20/news091.html

[2] https://ai-eng.doshisha.ac.jp/

[3] Tsuchiya S, Yoshimura E, Watabe H, Kawaoka T. The method of the emotion judgment based on an association mechanism. Journal of Natural Language Processing. 2007;**14**(3):119-238

[4] Tsuchiya S, Imono M, Yoshimura E, Watabe H. Emotion Judgment Method from a Meaning of an Utterance Sentence. Springer LNAI, No. 6881; 2011. pp. 367-376

[5] Tsuchiya S, Suzuki M, Imono M, Yoshimura E, Watabe H. Emotion judgement method based on knowledge base and association mechanism for colloquial expression. Transactions of the Japanese Society for Artificial Intelligence. 2014;**29**(1):11-20

[6] Watabe H, Horiguchi A, Kawaoka T. A sense retrieving method from a noun for the commonsense feeling judgement system. Journal of Artificial Intelligence. 2004;**19**(2):73-82

[7] Hirose T, Watabe H, Kawaoka T. Automatic refinement method of conceptbase considering the rule between concepts and frequency of appearance as an attribute. Technical Report of the Institute of Electronics, Information and Communication Engineers, NLC2001-93. 2002. pp. 109-116

[8] Kojima K, Watabe H, Kawaoka T. A method of a concept-base construction for an association system: Deciding attribute weights based on the degree of attribute reliability. Journal of Natural Language Processing. 2002;**9**(5):93-110

[9] Watabe H, Kawaoka T. Measuring degree of association between concepts for commonsense judgements. Journal of Natural Language Processing. 2001;**8**(2):39-54

[10] Mikolov T, Chen K, Corrado G, Dean J. Efficient estimation of word representations in vector space. CoRR. 2013;abs/1301.3781:1-12

[11] Sera T, Tsuchiya S, Watabe H. The method of measuring the degree of association between concepts using concept-base and Word2Vec. IEICE Technical Report, AI2017-50. 2018. pp. 43-48

Chapter 4

Utterance Emotion Estimation by Using Feature of Syntactic Pattern

Kazuyuki Matsumoto

Abstract

Emotion has been defined as basic emotions by various researchers, however, there are not many studies describing the relation between emotion and language patterns in detail based on statistical information. There are various languages all over the world, and even a language of the same country has different writing styles/expressions depending on which language media is used or who is a writer/speaker, which is thought to make it difficult to analyze the relation of emotion and language patterns. The author has been engaged in constructing and analyzing emotion corpora in some domains based on different sources. From the analysis results, emotion expressions started to become more understood that they have differences and tendencies according to the attributes of the writers and the speakers. In this chapter, I focused on the differences detected in the attributes of the writer/speaker with respect to language patterns; in usage tendencies or combinations of words, unknown expressions (slangs), sentence patterns, non-verbal expressions (emoji, emoticon, etc.) with relevant emotions, then introduce the outcome of the analytical survey on a large scale corpus obtained from a social networking service.

Keywords: emotion estimation, kansei robotics, slang, emoji

1. Introduction

In the research field of psychology, cognitive linguistics, it has been analyzed and studied about emotion and language [1, 2]. With regard to the relation between basic emotions and language, Fischer [3] performed cluster analysis and created a systematic chart based on emotion categories (emotion word) that can be expressed by language.

In the field of natural language processing, especially, sentiment analysis, a lot of researchers have been engaged in a study on the relationship between language patterns and emotion [4–8]. However, there are various languages all over the world, and language pattern varies depending on language media or writers. For this reason, there are no dictionaries describing language patterns and emotion cyclopaedically.

In the studies by Matsumoto [5] and Tokuhisa [9], they related language pattern dictionaries and occurred emotions. Mera et al. [10] proposed a framework to calculate degrees of positive/negative by using an emotion calculation formula for each case frame pattern. Because most of the methods proposed in these studies were assumed to be applied to "ideal" and "grammatical" sentences, they might not be effective for sentences on Internet.

Matsumoto et al. [11] proposed a method to estimate emotion in utterances including grammatically incorrect expressions such as Internet slangs. In the case of such casual expressions, it is thought to be more effective to take a method by machine learning based on a large scale natural language corpus than to register the knowledge into a dictionary. However, it is difficult to obtain a large scale corpus with labels, and it costs high to make such a corpus. Matsumoto et al. proposed a method to extract features based on word distributed representations as a robust method for unknown expressions. Their method converts words into distributed representation vectors and quantizes them with unsupervised clustering. They demonstrated that the method is robust to unknown expressions compared to existing methods.

After describing emotion estimation methods based on: dictionary, pattern and corpus, we introduce such important elements in corpus-based emotion estimation as gender differences and use of emoji expressions. Then we propose a deep learning-based method that uses a syntactic pattern as a feature combining the corpus-based method and the pattern-based method.

Section 2 introduces the emotion expression dictionary used in our previous research, Section 3 describes the emotion estimation by sentence patterns, and Section 4 explains the corpus-based emotion estimation method. Section 5 analyses emotion estimation with elements of gender and emojis. Section 6 propose a method based on syntax patterns, and Section 7 summarizes this chapter.

2. Emotion expression dictionary

Dictionaries collecting emotion expressions or evaluation expressions already exist [12–14]. These dictionaries defined emotional kinds that can be expressed with the words or phrases as classification categories and are registered them words or phrases. WordNet-Affect is a database created by extending WordNet thesaurus (conceptual database). A part of the information registered in WordNet-Affect is shown in **Table 1**.

There is a study that converted WordNet-Affect into Japanese language [15]. The evaluation polarity dictionary and the Japanese appraisal evaluation expression dictionary are language resources available for reputation analysis or opinion analysis,

A-Labels	Examples
EMOTION	*noun anger#1, verb fear#1*
MOOD	*noun animosisy#1, adjective amiable#1*
TRAIT	*noun aggressiveness#1, adjective competitive#1*
COGNITIVE STATE	*noun confusion#2, adjective dazed#2*
PHYSICAL STATE	*noun illness#1, adjective all in#1*
HEDONIC SIGNAL	*noun hurt#3, noun suffering#4*
EMOTION-ELICITING SITUATION	*noun awkwardness#3, adjective out of danger#1*
EMOTIONAL RESPONSE	*noun cold sweat#1, verb tremble#2*
BEHAVIOR	*noun offense#1, adjective inhibited#1*
ATTITUDE	*noun intolerance#1, noun defensive#1*
SENSATION	*noun coldness#1, verb feel#3*

Table 1.
A-labels and corresponding example synsets.

and they include words with annotation of emotion polarity; positive/negative. To analyze emotion of a sentence written in Japanese, an emotion expression dictionary including Japanese emotion expressions is necessary. It is also necessary to correspond linguistic resources to each language for emotion analysis written in foreign language. Because a framework of linguistic resource might be different according to the kind of language, it is difficult to make a unified dictionary.

In the case of Japanese language, the "Emotion Expression Dictionary" by Nakamura [16] is often referred to and often used in the field of natural language processing. However, many of the expressions included in the emotion expression dictionary are written words appeared in novels, therefore, there are some expressions that are rarely used as colloquial expressions. The Emotion synonym dictionary [17] also includes a few colloquial expressions, listing up the expressions which are thought to be useful for writing novels, scenarios and dramatic dialogs. Currently, as there are no dictionaries that cover practical language expressions such as colloquial expressions, such expressions or patterns are usually extracted from linguistic corpora.

As representative databases with registration of sentence patterns related to emotion expressions, there are EDR electronic dictionary [18], GoiTaikei: A Japanese Lexicon [19], and Kyoto University Case Frame [20]. However, because these linguistic resources are focused on semantic relations, emotion information is not annotated to these databases.

Using dictionaries has an aspect that known knowledge defined by human can be effectively used, however, it is often insufficient when it comes to dealing with things that are greatly related to human sensibilities such as emotions. While some words or expressions always give us unchangeable meanings or impressions, others change their meanings or impressions with the times. For example, the fairness and common sense toward the attributes such as race, religion and gender have changed significantly between decades before and today, so that this issue has been often referred to as one of the problems of artificial intelligence in recent years. Also, as language itself changes, dictionaries need to be updated constantly. In the form of a Wikipedia dictionary, some errors or old information are corrected or updated by being exposed to many people on the Web. However, such descriptions in the Wikipedia dictionary are based on the sensibility of the majority of people, it may not be possible to appropriately estimate the emotions of people with different sensibilities, so there is a limit to emotion estimation with just dictionaries.

3. Relation of sentence patterns and emotion

This section explains the relation of sentence patterns and emotion from the viewpoint of natural language processing by introducing the studies by Matsumoto [5] and Tokuhisa [9]. Matsumoto et al. [5] focused on the emotion occurrence condition for each sentence pattern to estimate emotion in dialog. They also constructed a dictionary that was registered emotion expressions to consider emotion values of each word. The emotion values mean the strength level of expressing each emotion.

Their study used a sentence pattern database that was extended the emotion calculation formula proposed by Mera et al. [10]. However, because they targeted basic sentence patterns, the method has the same problem with the existing method such as lack of versatility and it is weak to spoken expressions. The "Japanese Lexicon" [19] introduces a sentence pattern of each word. In the example of "Crying, "the sentence patterns are:

- N1 ga N2 wo Warau (N1 laughs at N2)

N1 and N2 are nouns. The emotion expressed by the sentence can differ depending on the noun applicable to N1 and N2. Referring to the example sentence: "Jiro cries over his debt," "debt" generally has a negative image. However, the emotion generated in this sentence can be affected by the speaker's attitude to "Jiro." These patterns were necessary to be annotated rules manually. **Figure 1** shows the case frame pattern of "N1 *ga* N2 *de/ni Naku*."

The following table (**Table 2**) shows some examples of sentence patterns and emotion occurrence rules. These information are saved as XML format on account of readability. **Figure 2** shows the emotion occurrence event sentence pattern database with XML format.

Matsumoto et al. [21] also extracted emotion occurrence event sentence patterns from a corpus. The following describes a flow of automatic extraction by Matsumoto et al. showing by example.

Step 1. The inputted sentence is analyzed by dependency parser. "CaboCha [22]"was used as the dependency parser.

First, according to the result of dependency parsing the last segment of the sentence is judged as a predicate. When a segment relates to the predicate and the end of the segment is either case particle or binding particle of "*ga*," "*ha*," "*wo*," "*ni*," "*he*," "*de*," "*to*," "*kara*," "*made*" or "*yori*" is extracted as surface case.

Step 2. The noun included in the obtained surface case element is annotated the semantic attributes based on "A Japanese Lexicon."

If the semantic attributes of the noun cannot be obtained, the basic form of the noun will be set into the surface case slot without annotating semantic attributes. The segment independent from the segment of predicate is not judged as case

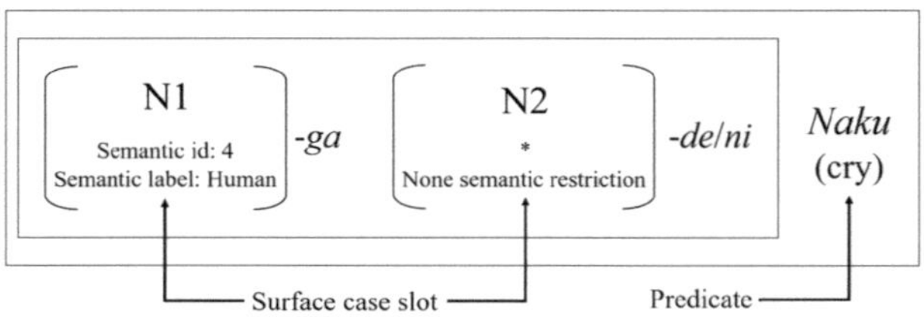

Figure 1.
Case frame pattern of "N1 ga N2 de/ni Naku".

Sentence Patern			*Sbj*	*FVN*	*EA*
English	**Case Pattern**	**Predicate**			
N1 cries over N2	N1[3]-*ga*, N2[*]-*ni/de*	*Naku*	N1	$f_{N2} \leq 0$	Sorrow
				$f_{N2} > 0$	Joy
N1 is angry at N2	N1[4]-*ga*, N2[*]-*wo*	*Okoru*	N1	N/A	Anger
N1 laughts at N2	N1[4]-*ga*, N2[*]-*wo*	*Warau*	N1	$f_{N2} \geq 0$	Joy
				$f_{N2} < 0$	Contempt
N1 worries about N2	N1[3]-*ga*, N2[*]-*wo*	*Ureeru*	N1	N/A	Anxiety
N1 is flustered by N2	N1[4]-*ga*, N2[1000]-de, *Ochitsuki-wo*	*Ushinau*	N1	N/A	Surprise
N1 subdues N2's pride	N1[4]-*ga*, N2[4]-*no*, *Hana-wo*	*Oru*	N2	N/A	Shame
N1 discommodes N2	N1[3]-*ga*, N2[4]-*ni*, *Meiwaku-wo*	*Kakeru*	N2	N/A	Hate
N2 is filled with N1	N1[3]-*ga*, N2[4,41,238]-*ni*	*Komiageru*	N2	N/A	N1's *EA*

Table 2.
Example of sentence pattern and emotion occurrence rule.

```
  <!-- 感情文型パタン辞書ファイル -->
▼<ESPatternDic name="Jepattern" jname="日本語感情文型パタン">
  ▼<EptItem id="0">
      <pred>愛読|する</pred>
      <read>アイドク|スル</read>
    ▼<emotionInfo>
        <emtag level="3.00" type="main">like</emtag>
      </emotionInfo>
      <image>3</image>
    ▼<e_pattern>
      ▼<case id="0" symbol="N1">
          <type>が</type>
          <sem>4</sem>
        </case>
      ▼<case id="1" symbol="N2">
          <type>を</type>
          <sem>92,011,101,037</sem>
        </case>
      </e_pattern>
    </EptItem>
  ▼<EptItem id="1">
      <pred>愛読|する</pred>
      <read>アイドク|スル</read>
    ▼<emotionInfo>
        <emtag level="3.00" type="main">like</emtag>
      </emotionInfo>
      <image>3</image>
    ▼<e_pattern>
      ▼<case id="0" symbol="N1">
          <type>が</type>
          <sem>4</sem>
        </case>
      ▼<case id="1" symbol="N2">
          <type>を</type>
          <sem>4</sem>
        </case>
      </e_pattern>
    </EptItem>
```


Figure 2.
XML format of sentence pattern database.

element. Because such sentence segment might be important element for deciding emotion attributes, it is extracted as modifier element. The obtained sentence pattern will be as 'EPT.'

Step 3. The set of emotion attributes 'E' annotated to the inputted sentence is decided as emotion attribute of 'EPT.' The combinations of 'EPT' and 'E' obtained from Step1 to Step3 are registered to the emotion occurrence sentence pattern DB. **Figure 3** shows an example of extraction process when "*Watashi wa odoroki no amari me wo shirokuro saseta*." is inputted.

This study automatically extracted sentence patterns from the emotion labeled corpus, created and evaluated the sentence pattern database. As the result of the cross-validation experiments for eight kinds of emotion estimations from sentences expressing emotions based on the corpus-derived sentence pattern database, approx. 42% emotion estimation accuracy was obtained.

Tokuhisa et al. [23] statistically analyzed the valency pattern of each sentence pattern, and proposed a method for emotion inference. Tokuhisa et al. [24] constructed and evaluated the dialog corpus by annotating emotion tags focusing on facial expressions of characters from manga comics.

Their study mainly target the utterances in dialogs, the target data are utterances not by actual persons but by fictional persons. Although these data are simulated

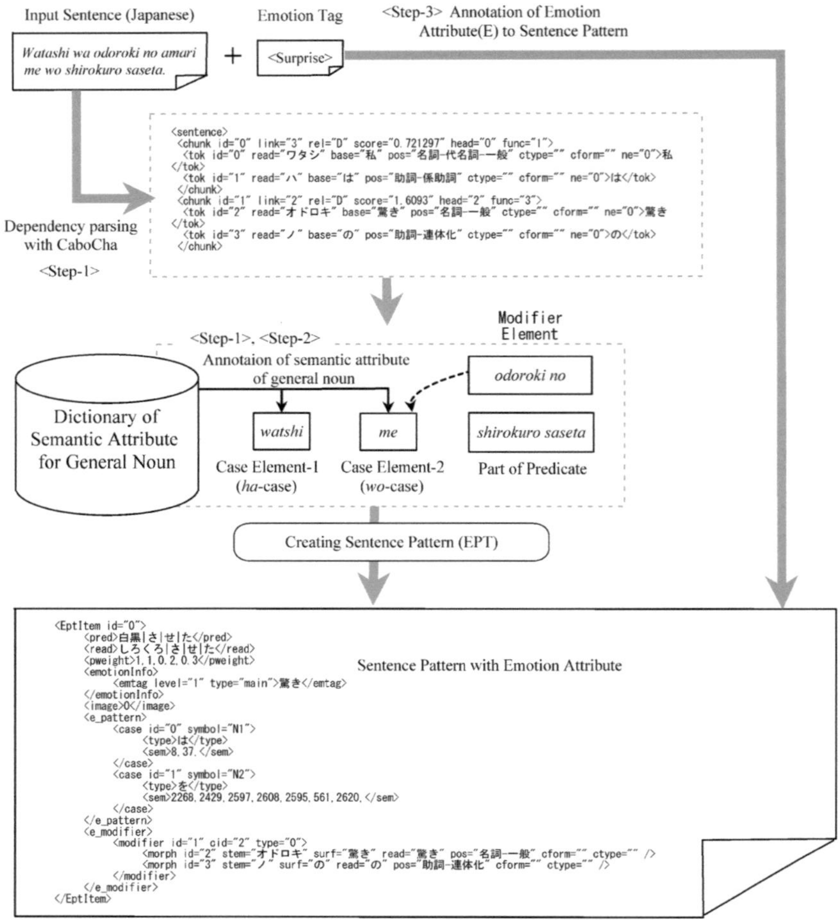

Figure 3.
*Flowchart of creation of sentence pattern (*EPT*).*

real dialog, it is considered that there exist some bias by the authors and generality might be lacking.

It is difficult to register entire colloquial expressions into a dictionary by strictly typifying their sentence patterns, besides, there are few challenging studies that try to annotate emotion that is subjective and sensitive to the sentence patterns. However, I thought that it would not be impossible to extract a relation between emotion and language patterns by studying thoroughly the recent corpus-based methods.

4. Corpus-based emotion analysis method

This section describes a corpus-based emotion analysis method by referring to the related literatures. The corpus annotated with emotion tags is defined as the emotion corpus. We would like to introduce existing studies that created and evaluated emotion analysis models based on statistical information and machine learning using emotion corpora.

4.1 Japanese-English parallel corpus [Minato et al.]

Minato et al. [25, 26] annotated emotion tags on word and sentence units included in Japanese and English parallel corpora. The completed corpus included

1,190 Japanese-English sentences. Based on the statistic results of the tagged words and sentences, they proposed and evaluated an emotion estimation method. They further considered the relevance between the two languages. Overview of their corpus is shown in **Table 3**.

The annotation to the corpus was made by the author, and evaluation by some examinees were not conducted. Matsumoto et al. [27] conducted an questionnaire on this corpus to several examinees and analyzed precision and recall between the tags annotated by the author and the tags annotated by some examinees. Because all examinees were Japanese people, they evaluated only Japanese sentences (1190 sentences). They calculated reliability of the annotation of the emotion tag by multi annotators. Reliability of tag annotation was calculated based on the match frequency among the three operators (initial tag annotator and two examinees). In their study, they proposed a method to reconstruct an emotion corpus by annotating reliability values. Reliability of tag is calculated with the Eq. (1). $\sum W(Tag_x)$ shows the sum of the weight of the tags annotated by the corpus creator and the two examinees

$$\text{Reliability}(Tag_x) = \sum W(Tag_x) \times \left(\frac{\text{Number of Matched Evaluators}}{\text{Total Number of Evaluators}}\right)^2 \tag{1}$$

They calculated the importance for each emotion category by calculating reliability of tag annotation. For calculation they used the weight of emotion tags according to the reliability as weight for emotion category instead of using simple word frequency. The calculation is based on the TFIDF method. Eq. (2) shows the weight of emotion category.

$$W^i_j = \alpha_i \times \sum RW^i_j \times \log \frac{N}{cf_j} \tag{2}$$

$$\alpha_i = \frac{1}{\sqrt{\sum_{m=1}^{l} \left(\frac{\sum RW^i_j}{cf_m}\right)^2}} \tag{3}$$

$\sum RW^i_j$ shows the sum of weight at emotion category 'E_i' in corpus. The 'cf_m' shows the number of emotion category tagged all sentences which included the word

Description	Frequency(J/E)	
Total # of Sentences	1190	1190
Total # of Words (Unique)	14202(2131)	11235(2409)
Average Words per Sentence	11.93	9.44
Total # of Emotion Sentences	601	601
Total # of Emotion Words (Unique)	1220(610)	1249(894)
Total # of Emotion Idioms (Unique)	274(231)	259(248)
Total # of Modifier Words (Unique)	108(70)	39(35)
Total # of Negative Words (Unique)	88(26)	31(15)
Average of Emotion Words/Idioms per Sentence	1.26	1.27
Average of times an emotion word did not carry emotion	0.22	0.15

Table 3.
Corpus statistics.

'w_m'. 'N' shows the total number of emotion category, and 'l' shows the total unique frequency of word. 'α_i' is normalization coefficient which is calculated with Eq. (3).

4.2 Corpus-based method using N-gram [Mishina et al.]

Mishina et al. [28] extracted word n-gram features from the emotion corpora, and proposed an emotion estimation method using the similarity score RECARE which was improved from BLEU often used for translation evaluation. The target emotion categories were four kinds; "anger", "joy", "hate", "hope". The problems of the method are; i) necessary to calculate similarity with all sentences in the corpus, and ii) the estimation accuracy affected by the corpus quality because the method is a simple example-based method.

4.3 Corpus creation and analysis [Quan et al.]

Quan et al. [29] constructed a large size of Chinese weblog emotion corpus "Ren-CECps," and analyzedthe corpus. In Ren-CECps, emotion tags were annotated to sentence, word, paragraph, and article units by some test subjects, and the corpus was analyzed from various viewpoints. The annotation to the corpus required hands, and as the size becomes larger and the corpus includes richer information, the higher annotation costs. There is a demerit that because the target are weblog articles, if there is bias in the writers, that will affect the quality of the corpus.

5. Analysis of emotion expressions according to gender

5.1 The emotion labeled corpus divided according to the users' attributes

In the study of Matsumoto et al. [30], they targeted the tweet sentences posted on Twitter and targeted each tweet for emotion estimation. Therefore, they needed to annotate emotion tags on each tweet sentence. The emotion estimation model was generated with the following steps:

1. The attribute labeled user account list is created from the accounts of popular users whose user attributes are known.

2. The tweets are collected for each user by using the attribute labeled user account list.

3. The four annotators manually annotate emotion tags on the collected tweets.

4. The emotion estimation model is created by extracting features from the tweets and by using a machine learning method.

In Step 4, the feature is extracted. First, the tweet sentence is split into word units by morphological analysis. Then, the words are converted into the distributed representations. They used another corpus to train the distributed representations. For about one year, they continued collecting tweets randomly; then, based on these tweets, they constructed a tweet corpus. They converted the corpus into the word-splitting format and used the text in this format for training the distributed representations.

Then, they annotated the emotion tags on the tweet sentences. Emotion tags annotated to the tweets are as follows:

- Positive emotions: "Joy," "Hope," "Love," "Relief," "Reception"
- Negative emotions: "Anger," "Hate," "Sorrow," "Fear," "Surprise," "Anxiety"
- Other emotions: "No emotion" and "Unclassified"

The total number of the emotion categories is 13. Some examples of the labeled tweets and their user attributes are shown in **Table 4**. The numbers of tweets for each emotion tag are shown in **Table 5**. As shown in **Table 5**, I found that there is bias in the numbers of tweets for each emotion.

In their chapter, they reported that emotion estimation accuracy increase by training the emotion corpus which is prepared for each attributes.

However, one thing to keep in mind when estimating emotion based on the corpus is who to annotate the corpus is. If the annotators' attributes and sensibilities are biased, a biased emotion estimation model would be built by learning the biased corpus. Such model cannot infer appropriate emotions according to the attributes of the authors or the speakers of the object sentence for estimation. To clarify the issue that attributes affect emotion estimation, the next subsection analyses what emotional expressions are used depending on gender based on the corpus.

5.2 Analysis of emotion expressions for each gender

In this section, I analyze emotion expressions by targeting on an emotion labeled corpus that are divided by gender. By investigating appearance frequency of emotion expressions included in the emotion expression dictionary and the kinds of the emotion labels annotated to the tweets including each emotion expression, I analyze appearance tendency of each expression according to gender by TF-ICF.

Emotion Tag	Tweet	Attribute	
		Sex	Job
Joy	At that time, we had a good match!	Male	Athlete
Hope	We did good! We should take a good rest and get together in good condition in the next live.	Male	Comedian
Sorrow	Coordinate plans for clothes are ruined by rain.	Female	Musician
Anxiety	All of the roads are in heavy traffic jam... I'm afraid if I could make it to the soccer game...	Female	Athlete

Table 4.
Labeled tweets and attributes of the users.

Joy	Hope	Love	Relief	Reception
6020	2135	487	67	15
Anger	**Hate**	**Sorrow**	**Fear**	**Surprise**
150	71	921	78	425
	Anxiety	**No emotion**	**Unknown**	
	301	943	46	

Table 5.
Annotation frequency of each emotion tag.

The formula of TF-ICF calculation is shown as Eq.(4) and Eq.(5). TF means Term Frequency, ICF means Inverse Category Frequency. TF_i^e shows word frequency appeared in emotion category *e*. ICF_i^e shows the value which is the number of emotion categories divided by the number of emotion categories including word *i*.

$$TF - ICF_i^e = TF_i^e \times ICF_i^e \tag{4}$$

$$ICF_i^e = \log \frac{|C|}{|\{C : t_i \in C\}|} \tag{5}$$

The TF-ICF calculation results for each gender are shown in **Table 6**. In this table, only top 10 expressions and TF-ICF scores are displayed.

From the analysis result, there are not significant difference between male and female. It is cause that only the general expressions are treated in the emotion expression dictionary for expression extraction.

	Joy					Sorrow			
	Male		Female			Male		Female	
Rank	Word	TF-ICF	Word	TF-ICF	Rank	Word	TF-ICF	Word	TF-ICF
1	*Minna*	309.4	*Minna*	383.3	1	*Konndo*	48.6	*Konndo*	48.6
2	*Ureshii*	188.8	*Ureshii*	323.0	2	*Itsuka*	19.5	*Kawaii*	31.1
3	*Shikkari*	153.7	*Daisuki*	190.7	3	*Kawaii*	17.5	*Itsuka*	31.1
4	*Chotto*	81.7	*Chotto*	118.7	4	*Wa*	17.5	*Mamonaku*	15.6
5	*Omoshiroi*	75.9	*Omoshiroi*	64.2	5	*Shimijimi*	17.5	*Yowamushi*	13.6
6	*Daisuki*	75.9	*Mina*	52.5	6	*Benn*	11.7	*Sugosu*	13.6
7	*Ooyorokobi*	56.4	*Zenzen*	48.6	7	*Tsumaranai*	9.7	*Benn*	9.7
8	*Deru*	52.5	*Zenbu*	42.8	8	*Mamonaku*	9.7	*Shizuka*	9.7
9	*Zenbu*	46.7	*Shikkari*	42.8	9	*Negau*	7.8	*Wa*	9.7
10	*Zenzen*	44.8	*Mattaku*	40.9	10	*Sugosu*	5.8	*Tsumetai*	7.8
	Love					**Anxiety**			
	Male		**Female**			**Male**		**Female**	
Rank	**Word**	**TF-ICF**	**Word**	**TF-ICF**	**Rank**	**Word**	**TF-ICF**	**Word**	**TF-ICF**
1	*Iru*	527.3	*Ii*	439.8	1	*Shi*	3712.8	*Shi*	3537.7
2	*Aru*	467.0	*Aru*	381.4	2	*Mashi*	2006.0	*Mashi*	1605.3
3	*Ii*	393.1	*Iru*	361.9	3	*Nai*	1109.0	*Nai*	1012.0
4	*Naru*	305.5	*Naru*	262.7	4	*Teru*	492.1	*Teru*	643.2
5	*Omou*	155.7	*Omou*	118.7	5	*Tanoshimi*	330.8	*Tanoshimi*	325.0
6	*Taisetsu*	70.1	*Taisetsu*	95.3	6	*Omoi*	321.6	*Ki*	235.5
7	*Hitsuyo*	37.0	*Kanjiru*	35.0	7	*Kore*	202.4	*Kore*	196.9
8	*Kanjiru*	27.2	*Atsui*	33.1	8	*Ki*	177.1	*Kansha*	151.8
9	*Atsui*	23.4	*Machi*	21.4	9	*Ganbari*	173.3	*Suki*	141.4
10	*Machi*	15.6	*Hitsuyo*	19.5	10	*Suki*	169.1	*Shiawase*	132.3

Table 6.
TF-ICF calculation results for each gender.

Joy				Sorrow			
Male		Female		Male		Female	
!	12.565865	!	12.496449	すみません	5.386413	さようなら	6.069605
^	12.001341		11.384511	申し訳	5.071520	悔し	6.019864
	10.443845	^	11.331308	残念	4.998112	メグ	4.512232
・	10.362423	‼	11.190660	相馬	4.815891	迷彩	4.512232
た	10.357401	📘	10.528542	配信	4.690341	・・・・・・・・	4.046404
の	10.321575	📙	10.528542	T	4.098353	イジグ	4.046404
て	9.947300	📗	10.528542	遅れ	3.664781	iPhone	4.027549
児嶋	9.903452	📕	10.528542	連携	3.611918	婆	3.714398
、	9.136461	#	10.465068	皮	3.611918	個	3.649186
飛	9.104408	の	10.240669	負け	3.587754	むくみ	3.611918
に	8.744838	た	10.208031	チャンス	3.583067	(´・(	3.501875
v	8.529375	て	10.022494	店員	3.501875	焼け	3.501875
弦	8.184875	^	9.891287	△	3.501875	ヘルペス	3.465736
。	7.917915	。	9.863882	誠に	3.501875	ごめん	3.337754
^	7.879624	:	9.731163	お祈り	3.501875	残念	3.301151
ます	7.809569	、	8.848594	動け	3.501875	悲しい	3.216726
まし	7.748093	に	8.566338	冥福	3.501875	痛	3.164503
~	7.747877	o	8.515148	亡くなり	3.501875	~(￣◇￣;)	3.092760
競馬	7.607281	୨୧	8.505089	潰し	3.501875	ディズニー	3.034803
#	7.522663	!	8.382803	喫し	3.465736	こぼし	3.034803
が	7.497108	っ	8.344959	痛く	3.452185	豆腐	3.015480
・	7.433959	絢香	8.331537	申し上げ	3.447335	02	3.008155
ありがとう	7.418473	`	8.317766	児嶋	3.301151	mayu	3.008155
で	7.351571	♡	8.261423	サーブ	3.243721	巡礼	3.008155
を	7.336861			痛み	3.143043	どい	3.008155
團	7.319106			マメ	3.143043	老け	3.008155
氣志	7.319106			悔しい	3.136576	剥け	3.008155
				ジャンケン	3.092760	久野	3.008155
				マイル	3.092760	ローブ	3.008155
				ラリー	3.092760	四国	3.008155

Anger				Anxiety			
Male		Female		Male		Female	
ヨー	4.512232	こら	3.865949	怖い	9.831017	予告編	3.611918
くたばれ	4.377344	パンティー	3.008155	つまらない	7.520387	怖い	3.454141
お嬢様	3.034803	被り	3.008155	顔面	4.512232	ストレス	3.034803
覚悟	3.008155	たれ	3.008155	怖	4.512232	渋滞	3.034803
例外	3.008155	出棺	3.008155	偏差	4.512232	下痢	3.008155
虫除け	3.008155	ラブホ	3.008155	闇	3.611918	ドンキ	3.008155
大江	2.626406	じん	2.407946	kunugifujihiko	3.501875	行為	3.008155
うそ	2.626406	たか	2.407946	眠い	3.243721	漏らし	3.008155
変態	2.626406	怒	2.407946	弦	3.015480	腫れ	2.876821
ひどい	2.626406	あやまる	2.407946	しだ	3.008155	心配	2.815267
うるさい	2.407946	シワ	2.407946	チャト	3.008155	地震	2.653141
しつこい	2.407946	(`	2.319570	ださ	3.008155	GP	2.626406
詐欺	2.319570	(- -#)	2.079442	千賀	3.008155	花粉	2.514435
つく	2.297662	尾行	2.023202	耳鳴り	3.008155	面	2.514435
運	2.122513	ドヤ	2.023202	地震	2.954859	恐怖＿	2.514435
パスタ	2.023202	ぶつぶつ	2.023202	怖かっ	2.626406	(;	2.462382
陰気	2.023202	マスコミ	2.023202	グチ	2.626406	怖く	2.462382
噛ま	2.023202	首都	2.023202	台風	2.486280	もしか	2.407946
眠たい	1.885826	ばか	2.023202	拾え	2.407946	しゅん	2.407946
blueriver	1.885826	526	1.969906	便乗	2.407946	未知	2.407946
蚊	1.885826	ayupon	1.969906	DM	2.407946	シワ	2.407946
バス停	1.750937	蚊	1.969906	不審	2.407946	ソワソワ	2.319570
アホ	1.750937	怒っ	1.750937	MEKURU	2.407946	こわ	2.319570
解放	1.750937	スキー	1.750937	乗り換え	2.407946	体調	2.299528
		装着	1.750937	付けよ	2.407946		

Table 7.
A part of TF-ICF calculation without limitation of emotional expression.

In addition, the results shown in **Table 7** were obtained by TF-ICF calculation without limitation of emotional expression.

It is found there are expressive differences of gender as seen from this result. For example, in both of gender, the symbols often be used in emotion: "Joy". Above all, female often use emoji. On the other hand, even though, comparatively, female use genial emotional expressions in emotion: "Anger", male often use radical expressions. It is considered that there are specific emotional expressions for each gender, and those express the gender difference of emotional expression.

Difference of emotional expressions by gender might decrease the estimation accuracy of the learned emotion estimation model due to gender bias. In order to avoid this, it would be useful to prepare an emotion estimation model for each gender or attribute, or to replace the expressions related to attributes with common expressions. In any case, it is clear that some sort of breakthrough is needed to maintain the fairness of machine learning.

5.3 Analysis of emoji

In this subsection, I analyze the appearance tendency of non-verbal expressions such as emoji according to gender. We analyzed usage trend of emoji from the total 59,009 tweets which were collected separately from the emotion corpus for each gender.

The results are shown in **Figures 4** and 5. In this figure, the horizontal axes shows Emoji type, the vertical axes shows frequency of use. In the graph of male, emojis with over 20 frequency are shown, and in the graph of female, emojis with over 100 frequency are shown. The types of emoji were set 4 classes; expression, emotion, exclamation and other. **Table 8** shows the result of emoji types and frequencies by counting emojis appeared over 10 times. As seen from this result, females had tendancy to use more emojis than male, and female often used emoji expressing expressions or emotions. As was expected that females would use more rich emotion expressions in their tweets, it was obvious from this usage trend of emoji. On the other hand, males used more exclamation marks than other types of emoji.

This result indicates that not only emotional expressions but also nonverbal expressions such as emojis have sufficient influence on emotion estimation. In addition to emojis, Japanese language has emoticons and ASCII art to convey various emotions. Globally, nonverbal expressions play important roles in

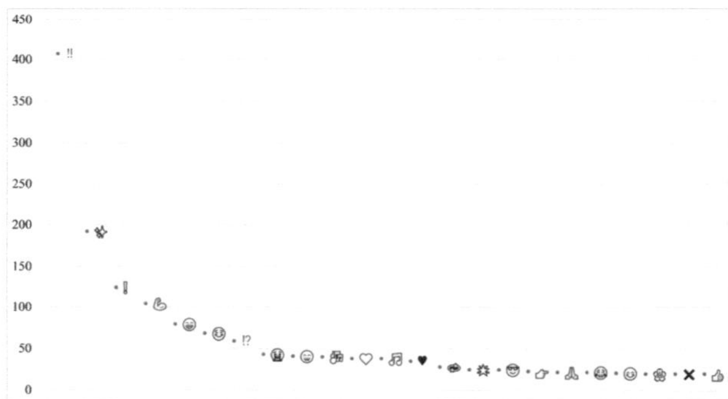

Figure 4.
Trend of emoji by male.

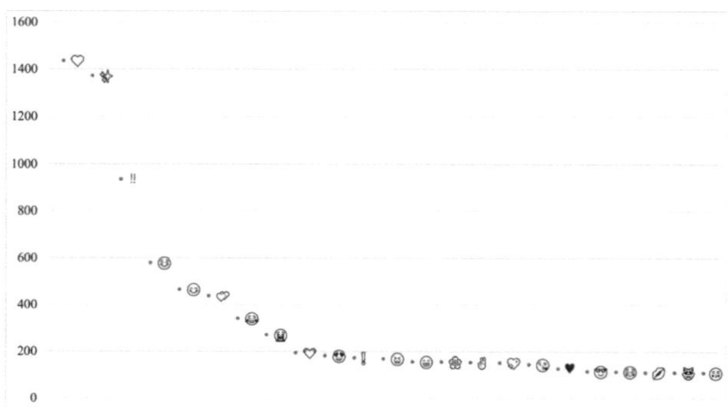

Figure 5.
Trend of emoji by female.

Gender	Type	Freq.	Rate
Male	Other	742	0.402
	Exclamation	600	0.325
	Expression	358	0.194
	Emotion	146	0.079
Female	Expression	3,536	0.32
	Other	3,312	0.3
	Emotion	3,002	0.272
	Exclamation	1,186	0.107

Table 8.
Emoji types and frequencies (over 10 times).

communication on the Web. From this, it is important to understand nonverbal expressions in order to estimate emotions.

6. Emotion estimation from feature of syntactic pattern by deep learning

6.1 Creation of emotion estimation by deep neural networks

We train language patterns that show emotions by using a deep learning method. We use syntactic patterns obtained from the parsing results by the Japanese dependency and case structure analyzer as features for learning. We use KNP [31] as the Japanese dependency and case structure analyzer. KNP is a syntactic, case and reference analyzer developed by Kyoto University. This system uses a noun case frame dictionary constructed by 7 billion web text.

As preprocessing of KNP, it is necessary to annotate morphological features on word unit by using a morphological analyzer. In this study, I make this annotation of morphological features by the morphological analyzer Juman [32]. As seen in **Figure 6**, sentences are analyzed by KNP.

As the result of analysis, the features are annotated on morpheme level and chunk level. The analysis result consists from "Clause layer", "Tag layer", "Morpheme layer". In this study, the features are extracted from the "Tag layer". For training, I use the features that have been annotated on chunk level to associate syntactic patterns with emotions. The examples of features annotated on chunk level are shown in **Table 9**.

The training data are the utterances annotated with emotion tags by manual. These utterances are used in the study by Matsumoto et al. [33], the source sentences are bilingual (Japanese-English). Because these sentences were used as

```
# S-ID:1 KNP:4.20-CF1.1 DATE:2021/03/09 SCORE:-11.50197
* 1D <文頭><連体修飾><用言:形><係:連格><レベル:B-><区切:0-5><ID:（形判連体）><連体節><状態述語><正規化
+ 1D <文頭><連体修飾><用言:形><係:連格><レベル:B-><区切:0-5><ID:（形判連体）><連体節><状態述語><正規化
素晴らしい すばらしい 素晴らしい 形容詞 3 * 0 イ形容詞イ段 19 基本形 2 "代表表記:素晴らしい/すばらしい
* -1D <文末><時制-過去><句点><体言><用言:判><レベル:C><区切:5-5><ID:（文末）><裸名詞><係:文末><提題受:
+ -1D <文末><時制-過去><句点><体言><用言:判><レベル:C><区切:5-5><ID:（文末）><裸名詞><係:文末><提題受:
景色 けしき 景色 名詞 6 普通名詞 1 * 0 * 0 "代表表記:景色/けしき カテゴリ:抽象物" <代表表記:景色/けしき
でした でした だ 判定詞 4 * 0 判定詞 25 デス列タ形 33 NIL <表現文末><かな漢字><ひらがな><活用語><付属>
。 。 。 特殊 1 句点 1 * 0 * 0 NIL <文末><英記号><記号><付属>
EOS
```

Figure 6.
Analysis results by KNP.

NE:ARTIFACT	NE:MONEY	Wikpedia hypernym: Film of USA	Counter:cm
NE:DATE	NE:ORGANIZATION	Wikipedia hypernym: Area	Counter:dollar
NE:LOCATION	NE:PERCENT	Wikipedia hypernym:Personal Name	Counter:meter
NE:MONEY	NE:PERSON	Wikipedia hypernym:Novel	Counter:person

Table 9.
Example of features.

examples for English composition, it is easy to extract syntactic patterns from sentences. As a preliminary experiment, I confirm emotion estimation accuracy by cross-validation. The breakdown of the five kinds of experimental corpora are shown in **Table 10**.

As the training, I use bi-directional LSTM (bi-LSTM) [34] which is extended LSTM (Long Short-Term Memory) [35]; a kind of recurrent neural networks. LSTM is suited to learning sequences. It enables efficient learning by memorizing and deleting past inputs. **Figure 7** shows the neural network structure using bi-LSTM. I use two LSTM layers.

In this study, I create a feature vector by chunk unit, and input the feature vector from the beginning of a sentence for training. The maximum number of chunks was set as 30 based on the maximum number of the chunks in the corpora.

Name	Cps1	Cps2	Cps3	Cps4	Cps5
# of Sentences	1190	1235	1554	1097	1054
# of Words	12548	8791	14980	20334	11860
# of Words per 1 Sentence	10.5	7.1	9.6	18.5	11.3
# of Vocabulary	2569	1753	2973	3890	2355
# of Clauses	6331	3789	6601	9231	5503
# of Clauses per 1 Sentence	5.3	3.1	4.2	8.4	5.2

Table 10.
Statistic of emotion tagged corpora.

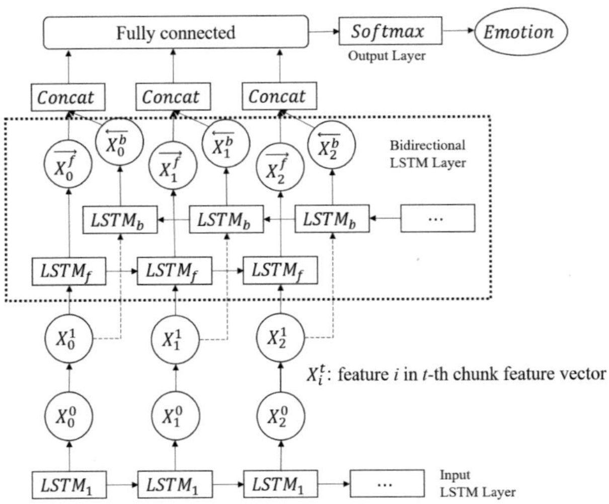

Figure 7.
Neural networks using bidirectional LSTM.

Emotion	Cps1	Cps2	Cps3	Cps4	Cps5
Joy	36.90%	65.50%	27.60%	59.50%	44.90%
Anger	43.10%	47.10%	19.40%	52.10%	38.40%
Sorrow	36.80%	41.20%	25.40%	32.00%	23.90%
Surprise	27.50%	58.50%	13.90%	24.60%	63.20%
Neutral	67.40%	8.70%	73.60%	46.00%	74.80%
Average	42.34%	44.20%	31.98%	42.84%	49.04%

Table 11.
F-measures of the preliminary experimental results.

Table 11 shows the result of the preliminary experiment. Averaged F-measure was 32–49%. The cause of this was thought to be the bias of emotion tags.

6.2 Experiment

I apply the emotion estimator trained syntactic features using bi-LSTM to the tweet sentences for each gender and evaluate the estimator by calculating accuracy. The architecture of the neural networks using bi-LSTM is shown in **Figure 8**. The tweet corpus shown in **Table 12** was used for the experiment.

We compare the result of the proposed method and the emotion estimation result based on emoji. The dictionary registered emojis and their expressing emotions is constructed as the Emoji Emotion Dictionary. The emoji emotion vectors of the emojis that are not in the dictionary are estimated. Emoji emotion vector of each emoji is obtained by calculating similarity with the seed emojis included in the emoji emotion dictionary and by acquiring emotion categories and similarities of top 5 similar seed emojis. The cosine similarity between the emoji distributed representations is used as the similarity of emojis. Eq. (6), (7), (8) shows the calculation of an emoji emotion vector.

$$EV_{e_i} = \left(ew_{e_i}^1\, ew_{e_i}^2 \ldots ew_{e_i}^j \ldots ew_{e_i}^n\right) \tag{6}$$

$$\begin{aligned} EV_{avg} &= \frac{1}{|EM_{topN}|} \sum_{em_{e_i} \in EM_{topN}} \left(sim_{e_i} \times EV_{e_i}\right) \\ &= \left(ew_{avg}^1\, ew_{avg}^2 \ldots ew_{avg}^j \ldots ew_{avg}^n\right) \end{aligned} \tag{7}$$

$$emotion = \arg\max_x ew_{avg}^x \tag{8}$$

Eq.(6) shows emotion vector EV_{e_i} of emoji em_{e_i}. Emoji emotion vector is a weighted mean of the emotion vectors of the top N similar seed emojis using similarity sim_{e_i} with seed emojis. $ew_{e_i}^j$ shows the weight of emotion category j. Eq.(7) is the formula to calculate the mean emoji emotion vector from the top similar N emoji set EM_{topN}. The estimated emotion is outputted as the emotion category x with the maximum weight value ew_{avg}^j of the mean vector by Eq.(8). The averaged emoji emotion vector is outputted by calculating emojis including in the sentences as the emotion estimation result. In this study, N value is set as 5 to estimate emotions.

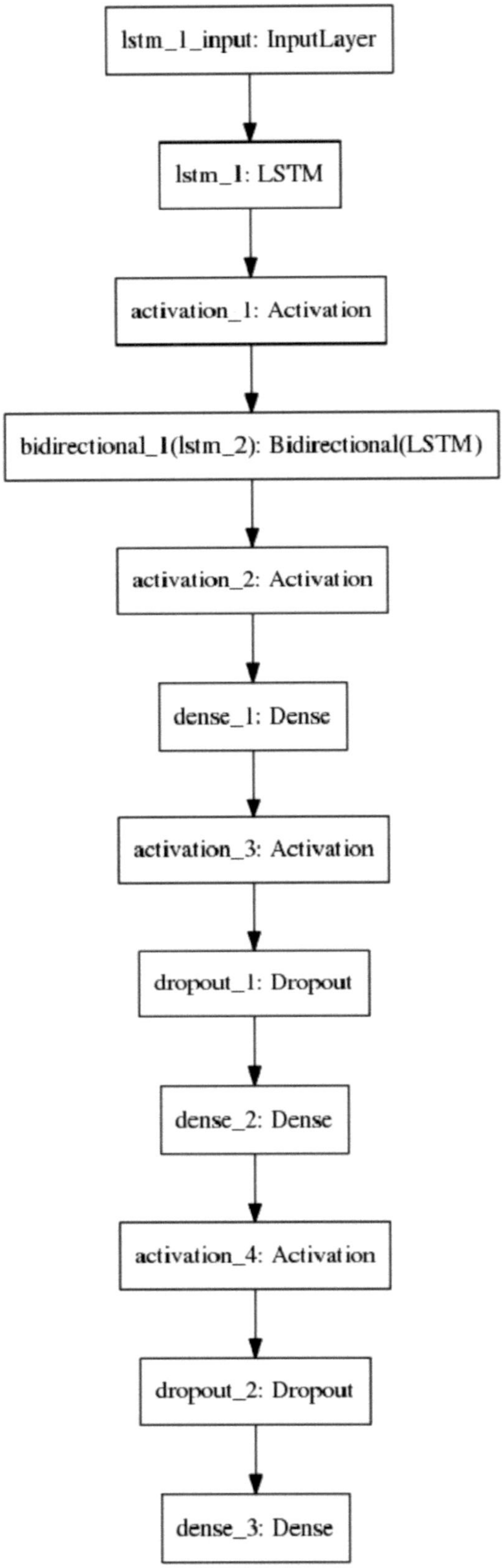

Figure 8.
bi-LSTM neural networks architecture.

	Joy	Anger	Sorrow	Surprise
Number of Sentences	995	418	997	219
Avg. Number of Clauses	8.3	7.5	7.9	7.3
Total Number of Words (Uniq.)	18599 (4766)	7163 (2177)	16523 (4015)	3670 (1337)
Avg. Number of Words	18.7	17.1	16.6	16.8
Total Number of Emojis (Uniq.)	1979 (161)	869 (68)	1581 (93)	379 (46)
Avg. Number of Emojis	1.99	2.08	1.59	1.73

Table 12.
Number of tweet sentences for each emotion.

	Proposed Method	Emoji-based Method
Joy	54.3%	46.3%
Anger	24.3%	4.0%
Sorrow	56.1%	24.6%
Surprise	30.6%	100.0%
Average	41.3%	43.7%

Table 13.
Comparison between the accuracies of the proposed method and the emoji-based method.

6.3 Experimental results

Because neutral tags were not annotated to the target tweet corpus, the accuracies for 4 emotion categories were calculated: "Joy," "Anger," "Sorrow," "Surprise." The experimental result is shown in **Table 13**. The highest accuracy was found for "Sorrow, "and the second highest was for "Joy. "The lowest accuracy was 24.3% and that was obtained for "Anger".

On the other hand, the overall accuracy was 43.7% by the emoji-based method, which was better than by the bi-LSTM based proposed method. However, the accuracy for "Anger" was low; 4% although the accuracy for "Surprise" was 100%. The primal reason is that the varieties of "Surprise" seed emoji were smaller than other kinds of emotions. It is also because that the number of tweets expressing "Surprise" with emoji was relatively scarce.

This result shows that using the syntax pattern enables effective emotion estimation using deep learning even with a small amount of learning data. It is thought that a more accurate model can be realized by flexibly changing dictionary knowledge depending on the domain or the speaker of the target sentence.

7. Conclusions

This chapter introduced our study on "emotion analysis on Japanese language" in the research field of the existing natural language processing and linguistic resources. Most of the existing approaches tried to associate emotions and language patterns, however, if language patterns express different emotions depending on the words consisting of the sentences, the rules for millions of combinations must be described.

It will be effective to analyze emotions based on corpora by annotating emotions on the corpora. In this chapter, various features were annotated on sentences by using a syntactic parser and feature vectors were generated by clause unit. The emotions of the tweet sentences were estimated by training the features using bi-LSTM neural networks.

It was also shown that the capability to development emotions from language patterns by using "emoji" as non-verbal expression. From the experimental results, the emoji-based method was found to be effective to tweet sentences including emoji. Because the amount of the emotion labeled data is limited and the existing dictionary and corpus-based methods cannot cover emotion expressions that are colloquially and depended on users' attributes, improvement of estimation accuracy is limited. Because emojis are non-verbal emotion expressions that can be used for all users, and the emoji expressions are not depended on the kind of languages, it is a hopeful key of emotion analysis in future.

In addition, syntax pattern might not be correctly extracted from the casual sentences that are often seen in dialogs on SNS. In that case, general-purpose neural language models such as BERT [36] and GPT-3 [37] will be useful. Future developments in language models might eliminate the necessity of human-defined linguistic knowledge such as syntactic patterns, however, methods such as fine tuning are still effective to build emotional estimation models satisfying the needs of all the people from large data. In that case, dictionary knowledge and syntax patterns will play effective roles in improving accuracy and presenting the basis for judgment.

Acknowledgements

This work was partially supported by JSPS KAKENHI Grant Numbers JP20K12027.

Author details

Kazuyuki Matsumoto
Tokushima University, Tokushima, Japan

*Address all correspondence to: matumoto@tokushima-u.ac.jp

References

[1] Kircanski, K, Lieberman, MD, Craske MG. "Feelings into words: contributions of language to exposure therapy" Psychol Sci. 2012 Oct 1;23(10):1086–91. doi: 10.1177/0956797612443830. Epub 2012 Aug 16.

[2] Kuperman, V, Estes, Z, Brysbaert, M, Warriner, AB. "Emotion and language: Valence and arousal affect word recognition", J Exp Psychol Gen. 2014 Jun;143(3):1065–1081. doi: 10.1037/a0035669. Epub 2014 Feb 3.

[3] Fischer, KW, Shaver, P, Carnochan, P. "A skill approach to emotional development: From basic- to subordinate-category emotions. In W. Damon (Ed.), Child development today and tomorrow, pp.107–136, (1989).

[4] Matsumoto, K, Ren, F, Kuroiwa, S. "Emotion Estimation System Based on Emotion Occurrence Sentence Pattern," International Conference on Intelligent Computing (ICIC 2006): Computational Intelligence. pp 902–911, 2006.

[5] Matsumoto, K, Mishina, K, Ren, F, and Kuroiwa, S. "Emotion Estimation Algorithm based on Emotion Occurrence Sentence Pattern," Journal of Natural Language Processing. 2007; **14**(3):239-271, https://doi.org/10.5715/jnlp.14.3_239. (in Japanese)

[6] Žižka, J and Dařena, F. "Automatic Sentiment Analysis Using the Textual Pattern Content Similarity in Natural Language," TSD 2010, LNAI 6231, pp. 224–231, 2010.

[7] Han, Y and Kim, KK. "Sentiment analysis on social media using morphological sentence pattern model," 2017 IEEE 15th International Conference on Software Engineering Research, Management and Applications (SERA), INSPEC Accession Number: 17009997.

[8] Su, S and Lynn, KT. "Extracting Product Features and Opinion Words Using Pattern Knowledge in Customer Reviews," The ScientificWorld Journal Volume 2013, Article ID 394758, 5 pages http://dx.doi.org/10.1155/2013/394758.

[9] Tokuhisa, M, Murakami, J, Ikehara, S. "Construction and Evaluation of Text-Dialog Corpus with Emotion Tags Focusing on Facial Expression in Comics," Conference: Knowledge-Based Intelligent Information and Engineering Systems, 10th International Conference, KES 2006, Bournemouth, UK, October 9–11, 2006, Proceedings, Part III, DOI10.1007/11893011_91.

[10] Mera, K. and Ichimura, T. "Emotion Analyzing Method Using Physiological State," Proceedings of KES 2004: Knowledge-Based Intelligent Information and Engineering Systems, pp. 195–201.

[11] Matsumoto, K, Yoshida, M, Xiao, Q, Luo, X and Kita, K. "Emotion recognition for sentences with unknown expressions based on semantic similarity by using Bag of Concepts," 12th International Conference on Fuzzy Systems and Knowledge Discovery, FSKD 2015, pp. 1394–1399, 2015.

[12] WordNet Affect: http://wndomains.fbk.eu/wnaffect.html

[13] Kobayashi, N, Inui, K, Matsumoto, Y, Tateishi, K. "Collecting Evaluative Expressions for Opinion Extraction," Journal of Natural Language Processing 12(3), pp. 203–222, 2005.

[14] Sano, M. (2010) Japanese Dictionary of Appraisal —attitude— (JAppraisal Dictionary ver1.0), published by Gengo Shigen Kyokai.

[15] Torii, Y, Das, D, Bandyopadhyay, S, Okumura, M. "Developing Japanese WordNet Affect for Analyzing

Emotions," Proceedings of the 2nd Workshop on Computational Approaches to Subjectivity and Sentiment Analysis, ACL-HLT 2011, pages 80–86, 2011.

[16] Nakamura, A. 1993. Emotional Expression Dictionary (in Japanese). Tokyodo shuppan.

[17] Ackerman, A and Puglisi, B. "The Emotion Thesaurus: A Writer's Guide to Character Expression, " Createspace Independent Pub. 2012.

[18] EDR Electronic Dictionary Version 1 Technical Guide, EDR TR2–003, 1995.

[19] Goitaikei – A Japanese Lexicon (in Japanese), Ikehara, S., Miyazaki, M., Shirai, S., Yokoo, A., Nakaiwa, H., Ogura, K., Oyama, Y., and Hayashi, Y. 1997. Iwanami Shoten

[20] Kawahara, D and Kurohashi, S. "Case Frame Compilation from the Web using High-Performance Computing," In Proceedings of the 5th International Conference on Language Resources and Evaluation (LREC2006), 2006.

[21] Matsumoto, K, Minato, J, Tsuchiya, S and Ren, F. "An Algorithm for Extracting Emotion Occurrence Sentence Pattern from Corpus," Proceedings of 2008 AIAC Symposium, 109–123, Tokushima, Mar. 2008.

[22] Kudo, T and Matsumoto, Y. "Fast Methods for Kernel-Based Text Analysis," Proceedings of the 41st Annual Meeting of the Association for Computational Linguistics (ACL2003), pp. 24-31, 2003.

[23] Tokuhisa, M, Tanaka, T, Ikehara, S and Murakami, J. "Emotion Reasoning Based on Valency Patterns–a prototype annotation of causal relationships—" Human and Artificial Intelligence Systems — From Control to Autonomy, Proceedings of the 4th International Symposium on Human and Artificial Intelligence Systems (HART2004), International Series on Natural and Artificial Intelligence. 2004;**1**:534-539.

[24] Tokuhisa, M, Ikehara, S and Murakami, J. "Construction of Text-dialog Corpus with Emotion Tags Focusing on Facial Expression in Comics," Journal of Natural Language Processing. 2007;**14**(3):193-217. (in Japanese)

[25] Minato, J, Matsumoto K, Ren, F and Kuroiwa, S. "Corpus-based analysis of Japanese-English of emotional expressions." In Proceedings of IEEE International Conference on Natural Language Processing and Knowledge Enginerring, pp. 413–418, 2007.

[26] Minato, J, Matsumoto K, Ren, F, Tsuchiya, S and Kuroiwa, S. "Evaluation of emotion estimation methods based on statistic features of emotion tagged corpus." International Journal of Innovative Computing, Information and Control, Vol. 4, No. 8, pp. 1931–1941, 2008.

[27] Matsumoto, K, Minato, J, Tsuchiya, S and Ren, F. "An Approach for Evaluating Emotion Tagged Corpus," In Proceedings of IADIS International Conference Interfaces and Human Computer Interaction 2008, pp. 258–262, Amsterdam, Jul. 2008.

[28] Mishina, K, Tsuchiya, S, Suzuki, M and Ren, F. "An Improvement of Example-based Emotion Estimation Using Similarity between Sentence and each Corpus," Journal of Natural Language Processing. 2010;**17**(4):91-110.

[29] Quan, C and Ren, F. "Sentence Emotion Analysis and Recognition Based on Emotion Words Using Ren-CECps," International Journal of Advanced Intelligence. 2010;**2**(1):105-117.

[30] Matsumoto, K, Yoshida, M, Kita, K, Wu, Y and Iwasa, F. "Effect of Users Attribute on Emotion Estimation from

Twitter," Proceedings of the 2nd IEEE International Conference on Computer and Communications, Vol.3, pp. 1186-1190, Oct. 2016.

[31] Kawahara, D and Kurohashi, S. "A Fully-Lexicalized Probabilistic Model for Japanese Syntactic and Case Structure Analysis," In Proceedings of the Human Language Technology Conference of the North American Chapter of the Association for Computational Linguistics (HLT-NAACL2006), pp.176–183, 2006.

[32] Japanese Morphological Analyzer: http://nlp.ist.i.kyoto-u.ac.jp/index.php?JUMAN

[33] Matsumoto, K, Ren, F, Yoshida, M and Kita, K. "Refinement by Filtering Translation Candidates and Similarity Based Approach to Expand Emotion Tagged Corpus," Knowledge Discovery, Knowledge Engineering and Knowledge Management. Communications in Computer and Information Science (CCIS). 2017;**631**:260-280.

[34] Schuster, M and Kuldip KP. "Bidirectional recurrent neural networks." IEEE Transactions on Signal Processing. 1997;**45**(11):2673-2681.

[35] Hochreiter, S and Schmidhuber, J., "Long short-term memory," Neural Computation. 1997;**9**(8):1735-1780

[36] Devlin, Y, Chang, MW, Lee, K and Toutanova, K. "BERT: Pre-training of Deep Bidirectional Transformers for Language Understanding." In Proceedings of the 2019 Conference of the North American Chapter of the Association for Computational Linguistics: Human Language Technologies. 2019;**1**:4171-4186.

[37] Brown, T, Mann, B, Ryder, N, et al. "Language Models are Few-Shot Learners." In Proceedings of Advances in Neural Information Processing Systems 33 (NeurIPS 2020), 2020.

Section 3

Dialogue System for Nursing Robot

Chapter 5

Issues in the Development of Conversation Dialog for Humanoid Nursing Partner Robots in Long-Term Care

Tetsuya Tanioka, Feni Betriana, Ryuichi Tanioka, Yuki Oobayashi, Kazuyuki Matsumoto, Yoshihiro Kai, Misao Miyagawa and Rozzano Locsin

Abstract

The purpose of this chapter is to explore the issues of development of conversational dialog of robots for nursing, especially for long-term care, and to forecast humanoid nursing partner robots (HNRs) introduced into clinical practice. In order to satisfy the required performance of HNRs, it is important that anthropomorphic robots act with high-quality conversational dialogic functions. As for its hardware, by allowing independent range of action and degree of freedom, the burden of quality exerted in human-robot communication is reduced, thereby unburdening nurses and professional caregivers. Furthermore, it is critical to develop a friendlier type of robot by equipping it with non-verbal emotive expressions that older people can perceive. If these functions are conjoined, anthropomorphic intelligent robots will serve as possible instructors, particularly for rehabilitation and recreation activities of older people. In this way, more than ever before, the HNRs will play an active role in healthcare and in the welfare fields.

Keywords: issues, conversation, dialog, humanoid nursing partner robots (HNRs), long-term care

1. Introduction

The issue of healthcare demands for the increasing older adult population in Japan is a significant concern, and in other developed countries [1]. This concern is further affected by the decreasing number of healthcare workers who are also getting older, resulting in high turnover rates of healthcare workers [2–4]. In this situation, it is appropriate to consider the use of healthcare robots, which is increasingly recognized as the potential solution to meet care demands of older persons as well as of patients with mental illness [5].

"What are the prominent areas of concern to support older persons when using healthcare robots?" and "What are the barriers to introducing Humanoid Nursing partner Robots (HNRs) to hospitals or elderly institutions?" Similarly, another question may be, "Which of these types of robots are needed for nursing care, the anthropomorphic or non-anthropomorphic robots?"

Different technological requirements can dictate whether anthropomorphic or non-anthropomorphic robots are needed. For example, if the technological demand is for measuring blood pressure and body temperatures, an anthropomorphic machine may not be necessary. Currently, non-robotic technologies are detecting and retrieving this information with digital hand-held devices, not necessarily robots [6]. However, anthropomorphic robots may be necessary when a conversation is expected, particularly during a dialog with older persons while taking blood pressures and other vital signs, much like a human nurse does today. In addition, the following distinctive questions are asked, "What nursing care tasks can be programmed specifically for anthropomorphic nursing partner robots?" and "What are the core competencies that only nurses, and professional caregivers can do?" Reflecting on the aforementioned questions, it is essential to establish a field of Robot Nursing science, developed by nurse scientists from the perspective of a unique ontology of nursing, and designed from a foundation of robotics engineering, computer science, and nursing science. The expectation is to develop a knowledge base for robot nursing science as foundation for the practice of nursing that uniquely embraces the anthropomorphic robot realities, particularly in demanding precise conversational capabilities. These realizations were illuminated through the posited questions from which the answers may further the development of robot nursing science and its practice.

The aim of this chapter is to explore the issues concerning the development of dialog robots for nursing, especially for long-term care, and the prospects for introducing HNRs into nursing practice.

2. Development of robots that can compassionate conversations

2.1 Concerns regarding human-robot conversation capabilities

As an issue for humanoid robot verbalization, the robot voice should have an appropriate intonation, the speech speed, and the voice range that is easy for older persons to hear [7]. If a cute-looking robot utters a low-pitched voice similar to that of an adult male, the user may find it creepy [8]. Therefore, it is necessary to consider a humanoid robot with easy voices for older persons to hear and voices that have a sense of familiarity.

Challenges to developing robot-nursing science are realistic. These challenges highlight the necessity to promote research with the goal of systematizing technological competencies, ethical thinking, safety measures, and outcomes of using robots in nursing settings. With new devices and technologies developed by engineers, introduced and used in nursing care, robot nursing science can only develop within an ontology of nursing at its core. The growing reality of healthcare robot utility is perceived as nursing partners in practice. Human caring expressed as human-to-human relationships, and among nonhumans are the futuristic visioning of healthcare with humanoid robots as main protagonists.

2.2 Development of caring dialogical database of humanoid nursing partner robots and older persons

Full and effective use of robots by nurses and healthcare providers would lead to a better understanding of patients and their needs. Thus, it is necessary to develop a "Caring Dialog Database" for HNRs in order to enhance robot capabilities to know the patient/client, and to share the expressions of human-robot interactions in esthetic ways. Furthermore, it is important to develop a dialog pattern that allows humanoid robots to empathize with an older person [9]. The ability to empathize and to communicate accurate empathy is likely to enhance the older person's feeling cared for through HNR actions such as: 1) Listening attentively and accepting of older persons; 2) Knowing older persons intentionally; and 3) Establishing appropriate caring dialog.

2.3 Robotics and artificial intelligence

Robotics and Artificial Intelligence (AI) will become a predominant aspect of healthcare and in welfare settings. Human caring was based on a human-to-human relationship. However, in a nonhuman-to-human relationship in the case of HNRs, it is essential to consider what is required in the aspects of ethical concerns and human safety. Regarding redefinitions of nursing and its underlying beliefs, values, and assumptions, it is pertinent to understand the implications of AI and its role in HNR in healthcare. Thus, robotics, AI, with Natural Language Processing (NLP) will become a predominant aspect of healthcare and welfare settings, particularly among older persons [10].

The HNRs must perform appropriate empathic dialogs [9, 11] by accurately judging the person's expression, non-verbal, and language expressions using AI sensory tools [12].

For the emotional recognition and non-verbal output, the required functions include: 1) Recognizing users' facial expressions; 2) Matching the expressions with emotion database information; 3) Selecting appropriate expressions from emotion database; and 4) Conveying emotional expression by particular motion, for example, using flashing light, moving upper limb and head, etc.

Furthermore, a robot's recognition and verbal output for voices by other persons, e.g., older persons can include: 1) Subjects' voice recognition; 2) Text conversion by NLP; 3) Matching with the NLP database for appropriate response; 4) Speech synthesis, and 5) Vocalization.

In conversations of robot with older person, it is expected that robot can provide accurate empathic response according to the situation. If an older person said,

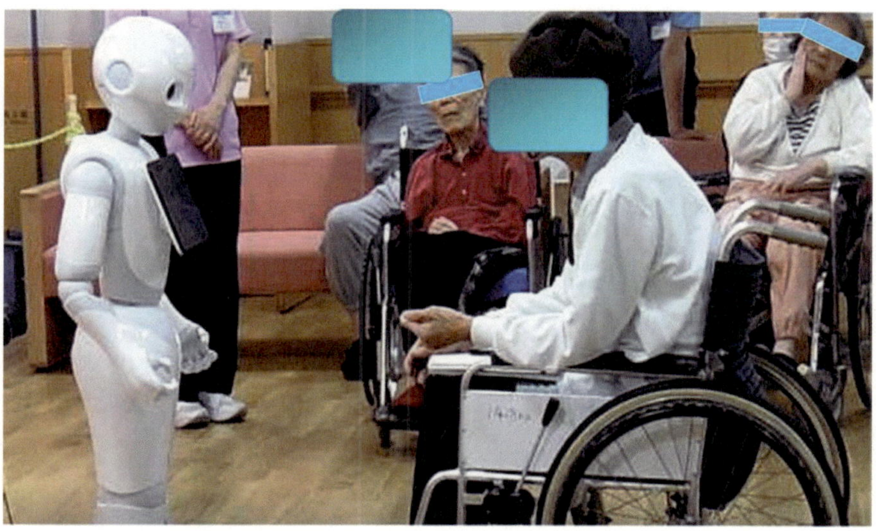

Figure 1.
Pepper is interacting with older persons.

"I want to eat sushi!", but the humanoid robot responds with, "I cannot eat because I am a robot", this is unlikely to engage older persons with the robot because such response does not demonstrate the empathic understanding of the robot. However, if the humanoid robot responds like so: "I am a robot, but I would like to try eating sushi. Tell me, what does it taste like?", this answer is likely to engage older persons because it relates a feeling of understanding and of empathy. If HNRs have this empathic response competency, older persons can attain well-being by understanding the content of dialogs and conversations with robots, such as Pepper robot (**Figure 1**).

3. Required conversation functions for humanoid nursing partner robots

This section discusses requirements for HNRs to allow a two-way conversation (dialog) with the user/other. HNRs should comprehend the content of the remarks, the intention of these remarks, including emotions, and others. From the information on speech, paralanguage, and appearance, such as facial expressions and gestures, HNRs can present listening postures to the user, using an appropriate 'line of sight' and nods to signify the appropriateness of the response to the user's remarks (**Figure 2**). These functionalities might facilitate active speech engagements of the user with the HNRs. Furthermore, when HNRs return appropriate responses based on the user's remarks and the contents they understand from non-verbal information, the user can feel that HNRs are listening and understanding dialog/conversation and may feel satisfied with the information or content of the interactive dialog.

As an example of an appropriate response by HNRs, a method such as repeating the keyword used by the user in the remark, or providing a topic related to the keyword can be considered. Further, the voice and movement of the HNRs during its response also affect the impression of the dialog (**Figure 3**).

For example, when a user speaks a sad topic or shows a sad expression, it is necessary that the HNR responds with a sentence and bodily behavior related to compassion, comfort, and encouragement. Moreover, it is important that the robot's voice is conveyed with a tone of artificial compassion that matches the response

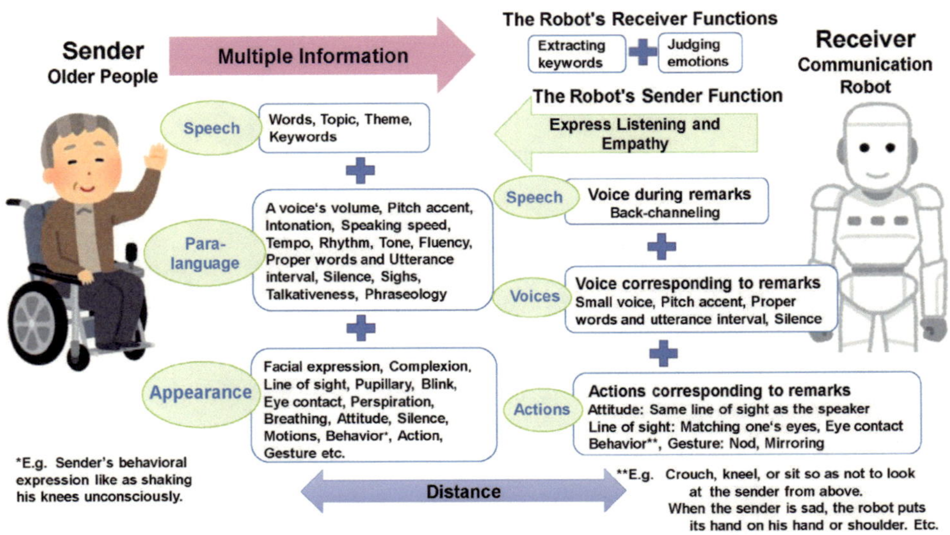

Figure 2.
An example of the required performance of the application for dialog/conversation with older persons (the robot's receiver function).

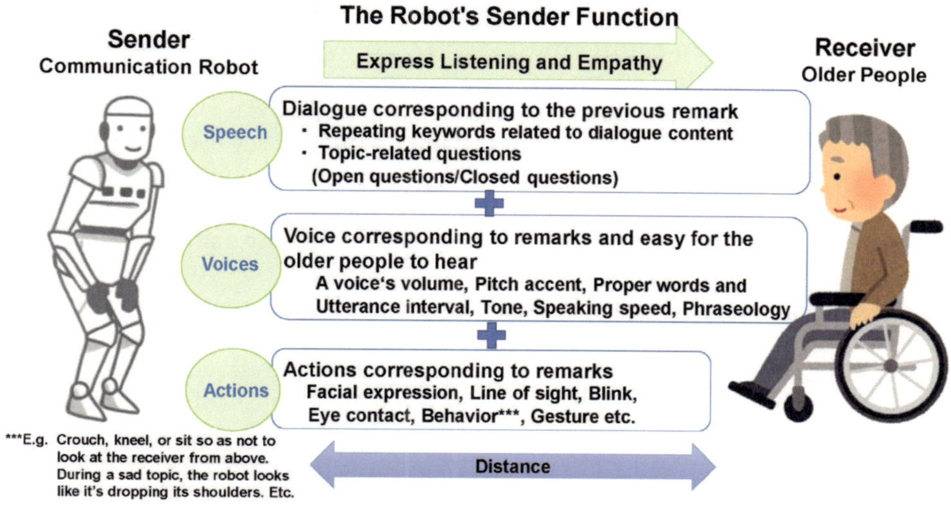

Figure 3.
An example of the required performance of the application for dialog/conversation with older persons (the robot's sender function).

sentence, and accurately delivers the humanoid robot's intention to the user [13, 14]. The humanoid robot's response matching the user's emotions may contribute to an expression of artificial sympathy and thereby enhancing empathic expressions for the user.

In the current facial expression recognition technology, reading facial expressions more accurately by recognizing a human face in 3 dimensions (3D) is also being studied [15, 16]. Research is being conducted on robot capabilities assessing human emotions not only from the movements of the eyes and mouth, but also from the movements of human facial muscles and upper body [17]. For the response ability of humanoid robots, it is at the stage wherein creation appropriations of response sentences are evidenced, including the examination of the response vocalization intervals without discomfort, and the implementations with robot verifications and improvement [18–21].

The required performance of the application for interacting with older persons requires a function that allows HNRs to speak according to the user's remarks, rather than the user conversing according to the robot's remarks. Alternatively, it is necessary to devise a way to give the user the feeling of having a dialog by making the user feel as if the HNR understands the user's remarks and intentions. As reception functions for the application, in addition to the technology to accurately read the content of remarks from the user's voice, the technology to read the user's situation from information other than voice such as facial expressions and gestures is required. Furthermore, as its transmission functions, a response sentence, vocalization and actions the expressive function matching the user's remarks and emotions are required.

4. How to develop a robot that can express verbal and non-verbal expressions

For a robot to convey verbal and non-verbal expressions like those of a human, it is necessary to have a receptor that is equivalent to that of a human. These are the sensory receptors for sight [22, 23], smell [24, 25], and touch [26]. Then, the information detected from these receptors is entered to the system. As such, it is necessary to perform machine learning based on the input information and prepare to output verbal and non-verbal expressions [27, 28]. Additionally, it is necessary for the robot to be able to perform the same movements and expressions as humans do in terms of its output [29, 30]. Depending on how similar the robot's expression is to human behavior, however, it may lead to an uncanny valley [31] that human beings may find creepy at some points, thereby influencing the responses it can do.

4.1 Anthropomorphic form necessary in human-robot conversation

The anthropomorphic form is necessary when an HNR is expected to talk to older persons or take vital signs like human nurse do. The influence of physiognomies of HNRs is a greater determining factor apparent to the efficient response of human beings in human-robot transactive engagements. Instead of the "Uncanny Valley" captivating robot communication, it is the human-HNR interactive 'fit' or congruence that may be better appreciated by human persons when HNRs are appropriately described from its appearance or looks. It's accurate and appropriate conversational capabilities further appropriate responses by HNR dependent largely on conversational communications that can easily be influenced by artificial affective communication [8].

In the case of a pleasant conversation with HNRs, human beings have a sense of affinity (Shinwa-kan) with HNRs, like appreciating their cuteness, and expressions of fun. However, human beings have also disappointed, especially when robots have poor conversational competencies. Human beings feel fear, misunderstanding, and confusion depending on the deviousness of the conversational language contents.

4.2 Roles and functions of humanoid nursing partner robots

Nevertheless, as companions in patient care, HNRs should assume multiple roles including being healthcare assistants to help with task completion. It is necessary for HNRs to possess abilities to express artificial emotions through linguistically appropriate and accurate communication processes, including nonverbal expressions with autonomous bodily movements. It is also critical that the appearance of HNRs would be more familiar, relatable, non-intimidating [32], does not cause human emotional unease and discomforts such as fear, anxiety, and suspiciousness, since human-like appearance of HNRs can lead to resistance [33].

One of the essential attributes of HNRs proposed is Artificial Affective Compassion (AAC) [8]. With the AAC (**Figure 4**) accentuating the significance

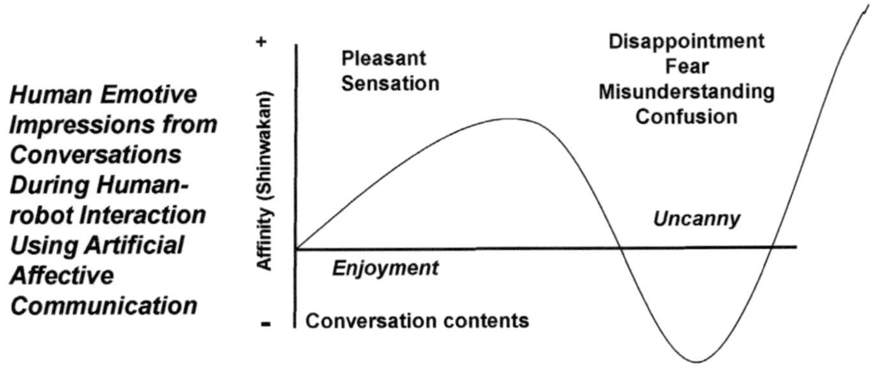

Figure 4.
Human Emotive Impressions from Conversations During Human-robot Interaction using Artificial Affective Communication (from Ref. [8]).

of language in human-robot interaction, not only will physiognomies of robot impact HNRs' value, but also its capabilities to communicate with AI for NLP. Communicating with artificial affection instilled with phonology and appropriately applied by mimicking human interactions through human features, elements designed with social and cultural nuances in communicative situations in transactive engagements with human beings may be made more valuable and meaningful for human healthcare practice.

5. An example of conversation with older persons and Pepper

The issue of conversation with Pepper includes expressions such as robot gaze [34], eye blink synchrony [35], eye contact [36], and speech [37]. This issue between older persons and Pepper with a conversation function in the application named "Kenkou-oukoku TALK for Pepper [38]" is vocalization with less intonation. This characteristic makes it difficult for older persons to understand whether Pepper's sentence is an interrogative or a declarative sentence. Similarly, it was found difficult for older persons to understand the end of the sentence with Pepper's talk. The pitch of Pepper's voice is high and difficult to hear. Similarly, Pepper's sensors may not be able to register the correct meaning of the sentence because of the older persons' soft voice or use of a dialect. If the contents of the conversation cannot be recognized, Pepper may interrupt the conversation or suddenly change the topic, which may offend older persons. Therefore, many situations exist wherein the contents of the dialog do not match. In the current performance of Pepper, it changed the topic while the user was still thinking about the answer to Pepper's question [39]. In addition, the operational issue of Pepper is its line of sight. If its line of sight is deviated from the talking person using Pepper's dialog program, Pepper will proceed with the conversation while recognizing others objects around it, thereby failing its line of sight. **Figure 5** presents Robot's line of sight.

The role of an intermediary to support the conversation between older persons and Pepper is important [40]. In the current conversation with Pepper, the user must adapt oneself to the Pepper's utterance. In this case, older persons are expected to listen to Pepper's talk instead of doing all the talking. They must have cognitive responsiveness ability while talking with Pepper. Training to respond quickly and accurately to Pepper's questions may be useful as a rehabilitation for cognitive function.

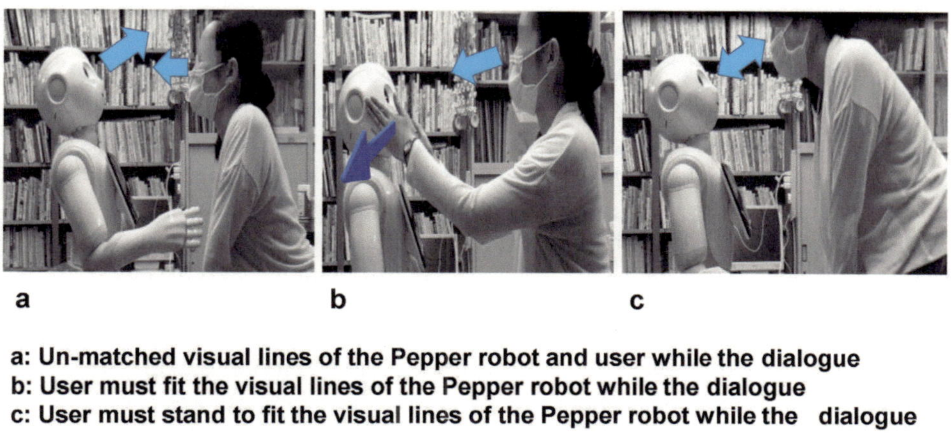

Figure 5.
Pepper's line of visualization.

Furthermore, to use the current Pepper's conversation application for cognitive rehabilitation of older persons, researchers propose a method in which older persons play a role of listeners. This role might be useful for training as they can concentrate on listening to the speaker's utterances, understand the content of the conversation, and to convey their personal feelings to the other person. When the conversation with Pepper is over, if the intermediary will instruct older persons to recollect the conversation content, this process may lead to maintenance of memory and confirmation, and training of information processing functions for older people.

As a means of improving the Pepper robot application, it is desirable that there is no one-way conversation by Pepper when older persons adapt the Pepper's utterances. Moreover, it is necessary to improve the conversation performance of the application so that older persons can enjoy talking with the robot for a long time. Thus, it is necessary to improve the following: (1) Timing of talk response; (2) Talk content must match the situation; (3) Appropriate reaction to the user's speech; (4) Functions of having eye contact with the user properly; and (5) Functions of reacting to users with non-verbal expressions. It is considered to assure the accomplishment of mutual conversation by these functions.

In order to solve the problem of line of vision, it is necessary to enable robots to express verbal and non-verbal expressions at the same level as human beings. It is considered that robots are merely showing artificial verbal and non-verbal expressions through machine learning [41]. Advanced intelligence is required when trying to express verbal and non-verbal expressions by incorporating artificial thinking, mind, and compassion [42]. Therefore, it is necessary to give the computer an artificial self [43].

Demands for quality nursing care and household responsibilities may be successfully met because of anticipated automation and robotization of work activities through AI and other technological advancements [44]. AI has become the latest "buzzword" in the industry today. To date, there is no AI machine able to 'learn' collective tacit knowledge. AI applies supervised learning and needs a great deal of data to do so. Humans learn in a 'self-supervised way'. Humans observe the world and figure out how it works. Humans need fewer data because humans can understand facts and interpret those using metaphors. Humans can transfer their abilities from one brain path to another. Moreover, these are skills, which AI will need if it is to progress to human intelligence [45].

6. Dialog systems

The types of dialog systems can be classified as follows: task-oriented dialog systems and non-task-oriented dialog systems. A task-oriented dialog system [46] performs the dialog necessary to achieve the demands of the user. The non-task-oriented dialog system [47] aims to continue the dialog itself. In order to continue the dialog, it is necessary to be able to handle non-task-oriented dialog [48]. The number of people who can converse at the same time is one robot to one person. However, in order to develop a high-performance humanoid robot in the future, it is desirable that one robot can have a dialog with three people.

Regarding the initiative of dialog, in the case of HNRs, the nurse and caregiver have the initiative. In the case of a dialog system, the distinction is made as to whether or not it has a physical body. For example, Siri does not have a body, but AI speakers and communication robots have a physical body. In this book, the HNRs having physicality are premised on the AI technology of mounted dialog processing.

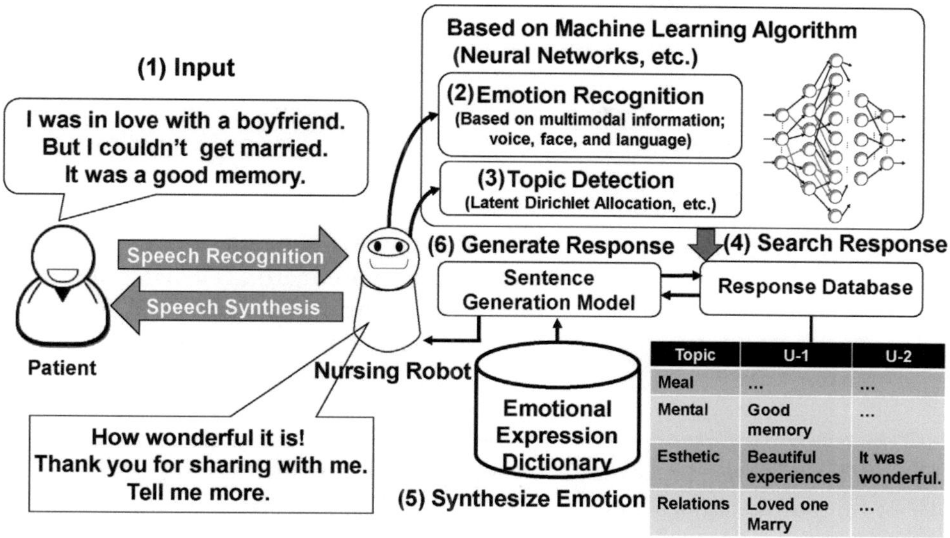

Figure 6.
Dialog processing mechanism using natural language processing.

As a classification of learning methods, regarding modality, in the case of a voice dialog robot, learning performed from a plurality of pieces of information (multimodal) [49] is needed. For example, information on a dialog between a skilled nurse and a care recipient is recorded at the same time. Then, it is necessary to let AI learn the motions and biological data acquired by the moving images and sensors as multimodal information.

The following are the steps involved in the HNR's generation of a response sentence containing emotions in response to the patient's speech. These steps are illustrated in **Figure 6**.

1. When the patient speaks, the HNR recognizes it as speech and converts it to linguistic information.

2. The HNR acquires multimodal information (vocal tone, facial expression, and language) from the patient via various mounted sensors and uses machine-learning algorithms, such as neural network algorithms, to estimate the patient's emotion [50].

3. Latent Dirichlet Allocation (LDA) [51] is applied to the patient's speech to detect the topics in their speech. The number of topics must be determined in advance. Here, the number of topics is determined as approximately 1000, and appropriate speech topics are acquired by clustering a set of topics obtained from the LDA based on the similarities between topics.

4. Based on the speech topic thus obtained and semantic features of the speech content, search and detect sentences from the response database built in advance that agree with the speech topic and have similar content to the speech. Obtain the most appropriate sentence in response to the speech content as the result.

5. Based on the response sentence thus obtained and the emotional estimated in (2), artificial emotions are synthesized using an emotion expression dictionary.

To synthesize emotions, find expressions from the emotion expression dictionary semantically similar to the expressions included in the response sentence and replace them with synonymous emotional expressions in accordance with the patient's emotions. If the patient has expressed a feeling of "sadness", then the expressions in the response sentence are replaced by expressions that suit the feeling of "sadness".

6. Lastly, a sentence is generated. Here, the response sentence is synthesized with artificial emotions and is corrected if it contains unnatural elements or if it does not match the context of the preceding sentence. Moreover, add some sentences (that evoke a speech response) so as to not stop the conversation with the patient.

This system requires speech act type estimation, topic estimation, concept extraction, and frame processing. Few studies have experimented and applied interactive processing using machine learning at a practical level in the field such as long-term care. It is considered that the cause is the difficulty of responding to situations that cannot occur in normal conversation, acquiring utterances, and predicting emotions in special dialog situations such as in long-term care.

In order to incorporate the tacit knowledge of nursing/long-term care into AI as explicit knowledge, we will record multimodal dialog data at the nursing/long-term care site. A multimodal nursing/long-term care corpus is constructed by its label to the data by skilled human nurses. Based on this corpus, we will develop a machine learning model method for predicting care labels from multimodal information. It is important to evaluate and tune the model with the goal of creating labels with high prediction accuracy. It might be an important point for future development to adapt the current mainstream method to dialog in long-term care with the older person and perform dialog processing. The following problems are considered when the target has a natural dialog with the dialog system only by adapting the current mainstream method.

- In a general dialog system, the next utterance cannot be made without the response of the other party, so the dialog may not continue depending on the situation.

- The general knowledge/common sense obtained from dictionary data such as Wikipedia may differ from the knowledge/common sense of the care recipient.

- Since there are cases where long-term care dialog is conducted based on tacit knowledge, there is a lot of information that cannot be seen from the collected corpus, and it is difficult for the machine-learning model to respond.

- Depending on the care recipient's situation, it may be difficult to predict emotions even with multimodal information.

We also believe it necessary to build a language model suitable for caregiving dialog with the older person for use with speech recognition, by collecting and analyzing a corpus of audio-visual caregiving dialogs. Thus, not only the language model but also the acoustic model must be tailored to the nursing practice and speech of older persons. In addition, by collecting and analyzing the audio-video corpus of long-term care dialog, it will be necessary to construct a language model suitable for elderly long-term care dialog for use in voice recognizers. It is thought that not only the language model but also the acoustic model that is suitable for the nursing care site and the voice of the older person will be required.

As HNRs learn to perform nursing functions, such as ambulation support, vital sign measurement, medication administration, and infectious disease protocols, the role of nurses in care delivery will change [52].

Hamstra [53] argued:

> *With less burden on nurses and improved quality of care for patients, collaboration with nurse robots will improve current trends of nursing shortages and unsafe patient ratios.*
>
> *While there is much to be excited about, there are aspects of nursing that cannot be replaced by robots. Nurses ability to understand the context, interpret hidden emotions, recognize implications, reflect empathy, and act on intuition, are innate and human skills that drive success as nurses.*

It is not obvious whether robotics can parallel these characteristics yet, as research on these topics is still ongoing. Also, it is important to remember that high-ability humanoid robots that can function much like a human being has not been developed in 2020. It is still in the developmental stage, and, today, it cannot be used as a functional work robot that is autonomous. Therefore, intermediaries such as healthcare providers play critical roles within the transactive relations between and among robot and older persons now.

Transactive is a term focusing on the transactional nature of things [54]. As an active process, it illuminates the main feature of the relationship among human-to-human and human-to-intelligent machines, which is, always a transaction. The term illuminates the relationship between HNRs and human persons. This picture shows a transactive relationship among older persons, an occupational therapist as intermediary, and Pepper (**Figure 7**).

Figure 7.
A transactive relationship among older persons, intermediary, and Pepper.

7. Prospects for the introduction of humanoid nursing partner robots in clinical nursing practice

Since AI and robots used for nursing and long-term care are diverse, it is important to clarify what AI is, what robots are used for in nursing and long-term care, how to use it, and how to apply it to nursing care? Therefore, conducting research related to this topic is important.

In addition, we will search for solutions in clinical settings used for nursing and clarify the required performance and required functions for AI. From the perspective of nursing and medical care, it is important to establish academic disciplines that explore and collaborate with AI and robot developers from the Faculty of Engineering. For example, consider how many and what kind of nursing work the healthcare robot can perform. Looking at the current situation, various robots that do not have human shapes have already "invaded" the medical world. There are robots for improving efficiency and accuracy, such as surgical robots and robots that support dispensing operations and assisting caregivers such as providing transfer and bathing support.

In the future, it can be imagined that a robot that scans the running of blood vessels and programs injection technology will perform injections. It may be possible to secure blood vessels for intravenous injection, which can be safer than nurses and doctors. However, what should we do to ensure safety, when this robot breaks down and out of control? Like humans, it is necessary to be able to judge whether to puncture anymore.

Will humans take on the role of monitoring robots in the relationship between robots and humans (nurses)? Will it add the role of monitoring robots to keep patients safe? Alternatively, now that computers have been introduced, will robot-dedicated engineers be stationed in hospitals just as computer engineers are stationed?

8. Conclusion

The purpose of this chapter is to explore the issues of development of conversational dialog of HNRs for nursing, especially for long-term care, and to forecast the robot introduction into clinical practice. The major issue is HNRs verbalization, which include inappropriate intonation, voice range, and the speech speed. These issues bring challenge to promote HNRs introduction into clinical practice. In order for robot to meet the demand and situations in nursing and healthcare, it is essential to improve conversation functions and performance of HNRs, such as the ability to express appropriate verbal and non-verbal expressions. For this challenge, collaboration between nursing researchers and AI and machine developers is recommended.

Acknowledgements

This work was partially supported by JSPS KAKENHI Grant Numbers JP17H01609. We would like to express our sincere appreciation to all the patients and participants who contributed to this article. With many thanks to Dr. Savina Schoenhofer for reviewing this chapter prior to its publication.

Conflict of interest

The authors have no conflicts of interest directly relevant to the content of this article.

Author details

Tetsuya Tanioka[1*], Feni Betriana[1], Ryuichi Tanioka[1], Yuki Oobayashi[1], Kazuyuki Matsumoto[1], Yoshihiro Kai[2], Misao Miyagawa[3] and Rozzano Locsin[1]

1 Tokushima University, Tokushima, Japan

2 Tokai University, Kanagawa, Japan

3 Tokushima Bunri University, Tokushima, Japan

*Address all correspondence to: tanioka.tetsuya@tokushima-u.ac.jp

References

[1] Muramatsu N, Akiyama H: Japan: super-aging society preparing for the future. Gerontologist. 2011;51(4):425-432. DOI: 10.1093/geront/gnr067

[2] Buchan J, Aiken L: Solving nursing shortages: a common priority. J Clin Nurs. 2008;17(24):3262-3268. DOI: 10.1111/j.1365-2702.2008.02636.x

[3] Murray MK: The nursing shortage. Past, present, and future. J Nurs Adm. 2002;32(2):79-84. DOI: 10.1097/00005110-200202000-00005

[4] World Health Organization. Global strategy on human resources for health: Workforce 2030 [Internet]. 2006. Available from: https://apps.who.int/iris/bitstream/handle/10665/250368/9789241511131-eng.pdf;jsessionid=F4F54C19AA58FCDA94DE34746F6DA886?sequence=1 [Accessed: 2020-11-24]

[5] Stuck RE, Rogers WA. Understanding older adult's perceptions of factors that support trust in human and robot care providers. In: Proceedings of the 10th International Conference on Pervasive Technologies Related to Assistive Environments, PETRA 2017; 21-23 June 2017; Island of Rhodes, Greece. New York, NY, USA: ACM, p.372-377; 2017. DOI: 10.1145/3056540.3076186

[6] Kakria P, Tripathi NK, Kitipawang P: A Real-Time Health Monitoring System for Remote Cardiac Patients Using Smartphone and Wearable Sensors. Int J Telemed Appl. 2015;373474. DOI: 10.1155/2015/373474

[7] Law M, Sutherland C, Ahn HS, et al.: Developing assistive robots for people with mild cognitive impairment and mild dementia: a qualitative study with older adults and experts in aged care. BMJ Open. 2019;9(9):e031937. DOI: 10.1136/bmjopen-2019-031937

[8] Betriana F, Osaka K, Matsumoto K, Tanioka T, Locsin RC: Relating Mori's Uncanny Valley in generating conversations with artificial affective communication and natural language processing. Nurs Philos. 2021;22(2):e12322. DOI: 10.1111/nup.12322

[9] Pepito JA, Ito H, Betriana F, Tanioka T, Locsin RC: Intelligent humanoid robots expressing artificial humanlike empathy in nursing situations. Nurs Philos. 2020;21:e12318. DOI: 10.1111/nup.12318

[10] Pou-Prom C, Raimondo S, Rudzicz F: A Conversational Robot for Older Adults with Alzheimer's Disease. ACM Trans Hum-Robot Interact. 2020;9(3):Article 21. DOI: 10.1145/3380785

[11] Nocentini O, Fiorini L, Acerbi G, Sorrentino A, Mancioppi G, Cavallo F: A Survey of Behavioral Models for Social Robots. Robotics. 2019,8(3),54. DOI: https://doi.org/10.3390/robotics8030054

[12] Behera A, Matthew P, Keidel A, et al.: Associating Facial Expressions and Upper-Body Gestures with Learning Tasks for Enhancing Intelligent Tutoring Systems. Int J Artif Intell Educ. 2020;30:236-270. DOI: 10.1007/s40593-020-00195-2

[13] Ray A. Compassionate Artificial Intelligence. Compassionate AI Lab; 2018. 258p. ISBN-10: 9382123466

[14] Kerasidou A: Artificial intelligence and the ongoing need for empathy, compassion and trust in healthcare. Bull World Health Organ. 2020;98(4):245-250. DOI: 10.2471/BLT.19.237198

[15] Tornincasa S, Vezzetti E, Moos S, et al.: 3D Facial Action Units and Expression Recognition using a Crisp Logic. Computer-Aided Design & Applications. 2019;16(2):256-268. DOI: 10.14733/cadaps.2019.256-268

[16] Agbolade O, Nazri A, Yaakob R, Ghani AA, Cheah YK: 3-Dimensional facial expression recognition in human using multi-points warping. BMC Bioinformatics. 2019;20(1):619. DOI: 10.1186/s12859-019-3153-2

[17] Vithanawasam TMW, Madhusanka BGDA: Dynamic Face and Upper-Body Emotion Recognition for Service Robots. Gaze and filled pause detection for smooth human-robot conversations. In: Proceedings of the 2018 IEEE/ACIS 17th International Conference on Computer and Information Science (ICIS); 6-8 June 2018; Singapore, Singapore; 2018, p.428-432. DOI: 10.1109/ICIS.2018.8466505

[18] Milhorat P, LalaEmail D: A Conversational Dialogue Manager for the Humanoid Robot ERICA. Advanced Social Interaction with Agents. 2018;119-131.

[19] Bilac M, Chamoux M, Lim A. Gaze and filled pause detection for smooth human-robot conversations. In Proceedings of the 2017 IEEE-RAS 17th International Conference on Humanoid Robotics; (Humanoids), 15-17 November 2017; Birmingham, UK; p.297-304. DOI: 10.1109/HUMANOIDS.2017.8246889

[20] Aoyagi S, Hirata K, Sato-Shimokawara E, Yamaguchi T. A Method to Obtain Seasonal Information for Smooth Communication Between Human and Chat Robot. In 2018 Joint 10th International Conference on Soft Computing and Intelligent Systems (SCIS) and 19th International Symposium on Advanced Intelligent Systems (ISIS), 2018; Toyama, Japan; p. 1121-1126. DOI: 10.1109/SCIS-ISIS.2018.00176

[21] Lala D, Nakamura S, Kawahara T. Analysis of Effect and Timing of Fillers in Natural Turn-Taking. In Proceedings of the Interspeech 2019, 15-19 September 2019; Graz, Austria; p. 4175-4179. DOI: 10.21437/Interspeech.2019-1527

[22] Simul NS, Ara NM, Islam MS. A support vector machine approach for real time vision based human robot interaction. In Proceedings of 19th International Conference on Computer and Information Technology (ICCIT), 2016; Dhaka, Bangladesh; p.496-500. DOI: 10.1109/ICCITECHN.2016.7860248

[23] Liu Z, Wu M, Cao W, et al.: A Facial Expression Emotion Recognition Based Human-robot Interaction System. IEEE/CAA Journal of Automatica Sinica. 2017; 4(4):668-676.

[24] Miwa H, Umetsu T, Takanishi A, Takanohu H. Human-like robot head that has olfactory sensation and facial color expression. In Proceedings 2001 ICRA. IEEE International Conference on Robotics and Automation; 21-26 May 2001; Seoul, South Korea. p.459-464. DOI: 10.1109/ROBOT.2001.932593

[25] Coradeschi S, Ishiguro H, Asada M, Shapiro SC, Thielscher M, Ishida H: Human-Inspired Robots. IEEE Intelligent Systems. 2006; 21(4):74-85. DOI: 10.1109/MIS.2006.72

[26] Martinez-Hernandez U, Prescott TJ. Expressive touch: Control of robot emotional expression by touch. In Proceedings of 2016 25th IEEE International Symposium on Robot and Human Interactive Communication (RO-MAN). 26-31 August 2016; New York. New York: IEEE; 2016. p.974-979. DOI: 10.1109/ROMAN.2016.7745227

[27] Li Y, Hashimoto M. Effect of emotional synchronization using facial expression recognition in human-robot communication. In Proceedings of 2011 IEEE International Conference on Robotics and Biomimetics. 7-11 December 2011; Karon Beach, Phuket. New York: IEEE; 2011. p.2872-2877. DOI: 10.1109/ROBIO.2011.6181741

[28] Yoon Y, Ko WR, Jang M, Lee J, Kim J, Lee G. Robots Learn Social Skills:

End-to-End Learning of Co-Speech Gesture Generation for Humanoid Robots. In Proceedings of the 2019 International Conference on Robotics and Automation (ICRA). 20-24 May 2019; Montreal; New York: IEEE; 2019. p.4303-4309. DOI: 10.1109/ICRA.2019.8793720

[29] Hua M, Shi F, Nan Y, Wang K, Chen H, Lian S. Towards More Realistic Human-Robot Conversation A Seq2Seq-based Body Gesture Interaction System. In Proceedings of the 2019 IEEE/RSJ International Conference on Intelligent Robots and Systems (IROS): 3-8 November 2019; Macau; New York: IEEE; 2019. p.247-252. DOI: 10.1109/IROS40897.2019.8968038

[30] Salem M, Rohlfing K, Kopp S, Joublin F. A friendly gesture: Investigating the effect of multimodal robot behavior in human-robot interaction. In 2011 RO-MAN. 31July-3 August 2011; Atlanta; New York: IEEE; 2011. p.247-252. DOI: 10.1109/ROMAN.2011.6005285

[31] Mori M: The Uncanny Valley. Energy. 1970;7:33-35 (in Japanese).

[32] Prakash A, Rogers WA: Why Some Humanoid Faces Are Perceived More Positively Than Others: Effects of Human-Likeness and Task. International Journal of Social Robotics. 2015;7(2):309-331. DOI:10.1007/s12369-014-0269-4

[33] Müller BCN, Gao X, Nijssen SRR, Damen TGE: I, Robot: How Human Appearance and Mind Attribution Relate to the Perceived Danger of Robots. International Journal of Social Robotics. 2020. DOI: 10.1007/s12369-020-00663-8

[34] Boucher JD, Pattacini U, Lelong A, et al.: I reach faster when I see you look: gaze effects in human-human and human-robot face-to-face cooperation. Frontiers in Neurorobotics. 2012;6(3). DOI: 10.3389/fnbot.2012.00003

[35] Tatsukawa K, Nakano T, Ishiguro H, Yoshikawa Y: Eyeblink synchrony in multimodal human-android interaction. Scientific Reports. 2016;6:39718. DOI: 10.1038/srep39718

[36] Xu T, Zhang H, Yu C: See you see me: the role of eye contact in multimodal human-robot interaction. ACM Trans Interact Intell Syst. 2016;6(1):2. DOI:10.1145/2882970

[37] Cid F, Moreno J, Bustos P, Núñez P: Muecas: a multi-sensor robotic head for affective human robot interaction and imitation. Sensors. 2014;14(5):7711-7737. DOI: 10.3390/s140507711

[38] XING INC. Kenkou-oukoku TALK for Pepper [Internet]. Available from: https://roboapp.joysound.com/talk/ [Accessed 2020-10-2]

[39] Miyagawa M, Yasuhara Y, Tanioka T, et al.: The Optimization of Humanoid Robot's Dialog in Improving Communication between Humanoid Robot and Older Adults. Intelligent Control and Automation. 2019;10(3):118-127. DOI: 10.4236/ica.2019.103008

[40] Osaka K, Sugimoto H, Tanioka T, et al.: Characteristics of a Transactive Phenomenon in Relationships among Older Adults with Dementia, Nurses as Intermediaries, and Communication Robot. Intelligent Control and Automation. 2017;8(2):111-125. DOI: 10.4236/ica.2017.82009

[41] Greeff J, Belpaeme T: Why Robots Should Be Social: Enhancing Machine Learning through Social Human-Robot Interaction. PLoS ONE. 2015;10(9):e0138061. DOI: 10.1371/journal.pone.0138061

[42] Asada M: Development of artificial empathy. Neuroscience Research. 2015;90:41-50. DOI: 10.1016/j.neures.2014.12.002

[43] Linda S: Endres, Personality engineering: Applying human personality theory to the design of artificial personalities. Advances in Human Factors/Ergonomics. 1995;20:477-482. DOI: 10.1016/S0921-2647(06)80262-5

[44] "Future of Work 2035: For Everyone to Shine" Panel. "Future of Work: 2035"-For Everyone to Shine-[Report] [Internet]. 2016. Available from: https://www.mhlw.go.jp/file/06-Seisakujouhou-12600000-Seisakutoukatsukan/0000152705.pdf. [Accessed 2020-11-30]

[45] Badimo, KH. How Artificial Intelligence can help to address some of the limitations of knowledge management. 2019. Available from: https://www.linkedin.com/pulse/how-artificial-intelligence-can-help-address-some-knowledge-badimo/ [Accessed 2020-11-30]

[46] Liao K, Liu Q, Wei Z, et al.: Task-oriented Dialogue System for Automatic Disease Diagnosis via Hierarchical Reinforcement Learning. ArXiv. 2020; arXiv preprint arXiv:2004.14254.

[47] Isoshima K, Hagiwara M. A Non-Task-Oriented Dialogue System Controlling the Utterance Length. In Proceedings-2018 Joint 10th International Conference on Soft Computing and Intelligent Systems and 19th International Symposium on Advanced Intelligent Systems (SCIS-ISIS 2018): 5-8 December 2018; Toyama, Japan. Institute of Electrical and Electronics Engineers Inc.; 2019. p.849-854. DOI: 10.1109/SCIS-ISIS.2018.00140

[48] Zhou Y, Black AW, Rudnicky AI. Learning Conversational Systems that Interleave Task and Non-Task Content. In: Proceedings of the Twenty-Sixth International Joint Conference on Artificial Intelligence (IJCAI-17): 19-25 August 2017; Melbourne, Australia. International Joint Conferences on Artificial Intelligence; 2017. p.4214-4220. https://www.ijcai.org/Proceedings/2017/0589.pdf

[49] Fernández-Rodicio E, Castro-González Á, Alonso-Martín F, Maroto-Gómez M, Salichs MÁ: Modelling Multimodal Dialogues for Social Robots Using Communicative Acts. Sensors. 2020;20(12):3440. DOI: 10.3390/s20123440

[50] Khalil RA, Jones E, Babar MI, Jan T, Zafar MH, Alhussain T: Speech Emotion Recognition Using Deep Learning Techniques: A Review. IEEE Access. 2019;7,117327-117345. DOI: 10.1109/ACCESS.2019.2936124

[51] Blei DM, Ng AY, Jordan MI: Latent Dirichlet Allocation. Journal of Machine Learning Research. 2003;3:993-1022.

[52] Robert N: How artificial intelligence is changing nursing. Nursing Management. 2019;50(9),30-39. DOI: 10.1097/01.NUMA.0000578988.56622.21

[53] Hamstra, B. Will these nurse robots take your job? Don't freak out just yet [Internet]. 2020. Available from: https://nurse.org/articles/nurse-robots-friend-or-foe/ [Accessed: 2020-12-23]

[54] Tanioka T: The Development of the Transactive Relationship Theory of Nursing (TRETON): A Nursing Engagement Model for Persons and Humanoid Nursing Robots. Int J Nurs Clin Pract. 2017;4:223. DOI: https://doi.org/10.15344/2394-4978/2017/223

Chapter 6

Robot Therapy Program for Patients with Dementia: Its Framework and Effectiveness

Kyoko Osaka, Ryuichi Tanioka, Feni Betriana, Tetsuya Tanioka, Yoshihiro Kai and Rozzano C. Locsin

Abstract

Robot therapy uses humanoid and animal-like robots. The robot therapy for older adults is expected to affect the therapeutic goals, including physical condition, cognitive function, and provide joy. By interaction with humanoid or animal-like robot, the older adults who are not physically active may have the improvement of their physical condition, such as hugging, stroking, talking with them, and participating in any activity involving the robot. The typical examples show that animal therapy has almost the same effectiveness as robot therapy among older people. It is clarified that robot therapy can be expected to have a healing effect on patients, improve motivation for activity, and increase the amount of activity, like animal therapy. Furthermore, it was essential to consider the intermediary role of nurses for connecting the robot and older adults and their role, even if the robot is not sophisticated enough to be useful as a humanoid nurse robot for rehabilitation and dialogue with older adults. Thus, robot therapy could be considered another important intervention in the challenging health and innovative care practices needed in the care of older persons. This chapter explains the robot therapy program for patients with dementia from the viewpoint of its framework and effectiveness.

Keywords: robot therapy, animal therapy, older adults, dementia, intermediary role

1. Introduction

In Japan, the number of older adults requiring medical and nursing care is increasing, constituting a super-aged society [1]. This trend is exacerbated by the decrease in the active working population and is accompanied by a declining birth rate [2]. This situation in Japan and other countries, including for current and future rehabilitation services, increases the demand for nursing care of older adults. With that as a result, nursing staff shortages are becoming more serious [3, 4]. Therefore, it is necessary to bridge the gap between human resources and demand for services in health care. In addition, the number of patients with dementia is also increasing especially among the older adult population who need more engaged medical and nursing care [5].

There are plenty of reports on the benefits of animal therapy, which began in the USA in the 1970s [6]. One study conducted in patient with schizophrenia found that showed cortisol level was significantly reduced after participating in an animal assisted therapy session, which could indicate that interaction with the therapy dogs reduced stress [7]. Another study reported that measuring actigraphy increased sleep duration (min) when visitors were accompanied by a dog rather than the robot seal or soft toy cat [8]. Another study reported that animal therapy is associated with decreased impulsivity, aggression, and anxiety, and increased sociability [9].

Meanwhile, several studies have suggested that the robots used in robot therapy can improve the cognitive level and reduce the Behavioral and Psychological Symptoms of Dementia (BPSD) in patients [10, 11]. However, none of the studies have tested the robots on a large sample, meaning that their findings have limited generalizability [12]. Yokoyama reported the caregiver must play an intermediary role during robot therapy for older people with dementia [13]. For such therapy to be effective, it must be evaluated from the perspective of the user (an older adult with dementia) and the caregiver (a nurse or other professional caregiver).

Osaka and other studies [14–16] analysed Heart Rate Variability (HRV) and accelerometer data in two-second increments and showed real-time results. As such, changes in autonomic activity and the intensity of physical exercise could be determined during the implementation of robot therapy. This device sensor was designed to be small size and thus carry only minimal burden for older persons. Another advantage was that it could transmit data to a computer wirelessly, allowing the tested subjects to move freely. It is also possible for an observer to supplement the data by recording through the participant observation of the relationship between the older person and the caregiver during the intervention.

Limited study on robot therapy in older adults with dementia has assessed the intervention objectively and comprehensively by using participant observation and HRV and accelerometer data. The study by Osaka et al. [14] is valuable in that it provides objective data to those involved in caring for older adults with dementia (caregivers and health care professionals involved in rehabilitation), thereby allowing them to review their interventional approaches in relation to standard. Moreover, if it can be demonstrated that low-cost robot therapy is effective, then the study will offer valuable data for developing policies on cost-effective robotics in dementia care.

This chapter explains the robot therapy program for patients with dementia from the viewpoint of its framework and effectiveness.

2. Definition of terms

2.1 Animal therapy

The American Veterinary Medical Association (AVMA) defined Animal Assisted Therapy (AAT) as one of the Animal Assisted Intervention (AAI) [17]. There are various animals are used for AAT, such as canines, felines, and equines, depending on purpose of treatment, but the most frequently used animal for AAT is the dog [18]. The AAI for older people can be expected to have the effect of suppressing the decline in cognitive functions and improving the peripheral symptoms (depression, agitation, aggression) and insomnia associated with dementia [10].

2.2 Robot therapy

The number of robot therapy articles for rehabilitation or recreation that include communication is increasing. Considering the entity of a care robot, several definitions are recently offered [19]. However, there is no consensus about their findings. The devices and applications in those studies have yet to be integrated into widespread clinical use [20]. Robot therapy functions include providing therapy, educate, enable communication, and so on [21]. A pilot study showed that by interacting with Paro, a seal-like robot, the communication and interaction skills, and activity participation of older people improved [22]. Another study reported that the use of Paro is associated with improvement in emotional state and social interaction and reduce the challenging behaviours among older people [23]. Research showed that robot therapy has the same effect on people as animal therapy [21]. The effects of robot therapy and animal therapy influence the physical, cognitive, and mental conditions of the users, especially the older people.

3. Theoretical framework

The Model for the intermediary Role of nurses in Transactive relationships with Healthcare robots (MIRTH©) [24] explains the engagement processes that are characteristic activities of older adults with dementia, the nurse as mediator, and the communication robot (**Figure 1**). Healthcare robots' function in transactive relationships among patients and nurses. The nurses' role as intermediaries is integral to facilitating the interaction between these robots and the older adult patients who are in transactive relationships. The effects of the intermediary role are especially prominent with low-fidelity robots in use today. The functional abilities of the nurse as intermediary include knowledge of advancing technologies regarding robots that foster quality care.

Nurses as intermediaries should: (1) have an accurate awareness of each of the functions of robot performance and the usefulness of each function relevant to patient care situations; (2) create relationships with healthcare robots so that they can promote the health and safety of older adults while increasing their enjoyment through physical and social activities; and (3) seek safe, secure, and competent ways to facilitate using healthcare robots for healthcare. In essence, intermediaries

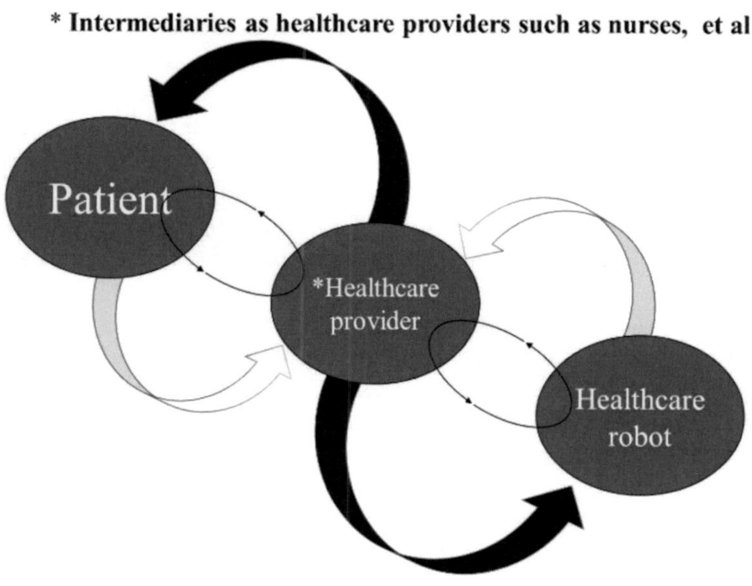

Figure 1.
Illustration of MIRTH©.

should prepare the environment for using healthcare robots. In doing so older adults can use healthcare robots for complicated operations, with the nurse as intermediary monitoring the effective use of robots and identifying clinical problems while working with healthcare institutions to address preventable healthcare problems.

The MIRTH© model has five assumptions:

It is the responsibility of nurses as professionals to practice nursing grounded in discipline-related knowledge of nursing. The most important attribute in nursing is the relationship expressed as caring. Robot performance requires an intermediary for their effective and safe use [14, 15]. This assumption expresses the importance of the functions of intermediaries in robot-human situations. The intermediary is inextricably linked with the patient and the healthcare robot;

- Robots are used for rehabilitation, recreation, and caring of older adults [16]. This assumption describes the variety of functions of robots specifically for older persons;

- High-quality care with robot-human relationship is guided by ethical and moral standards of nursing [25]. With human beings as patients and robots as integral to human health care, this relationship must be linked with considerations of beneficial effects founded on justice and goodness;

- Technologies of health and nursing are elements for caring [26].

- The utility of advancing technologies founded on competent expressions of caring provides opportunities for innovating human caring practices;

- Nursing is both a discipline and a profession [27].

It is the responsibility of nurses as professionals to practice nursing grounded in discipline-related knowledge of nursing. The interactive engagement, the lived experience of the caring between patients and nurses, gives meaning to the nursing relationship, the most important attribute in Nursing.

Framework for robot therapy program.

It has been reported that therapies using animals have a healing effect on patients and an improvement in motivation for performing activities [28, 29]. Park et al. performed a meta-analysis on animal assisted and pet robot interventions which suggested that AAI and Pet Robot Intervention (PRI) significantly reduce depression in patients with dementia. It report, nine studies were analysed and seven of them showed confirming results. The outcome measurements used scales such as functional tests and depression scales. In the two studies, pulse oximetry, pulse rate or galvanic skin response (GSR) (electric skin response) were combined and evaluated as physiological indicators [10].

Intervention therapies using animals for hospitalized patients is not uncommon. Studies by Osaka and others [14–16] suggest that robot therapy is expected to have a more healing effect on patients and improve motivation for activities for older people by using an expensive humanoid robot such as Pepper from an inexpensive communication robot.

Figure 2 shows the framework of the effectiveness using robot therapy by Osaka. Robot therapy uses humanoid and animal-like robots. The robot therapy is expected to affect the therapeutic goals, including physical effect (e.g., relaxation, motivation), physiological effect (e.g., improvement of vital signs), and social effect (e.g., stimulation of communication among inpatients and caregivers) [30].

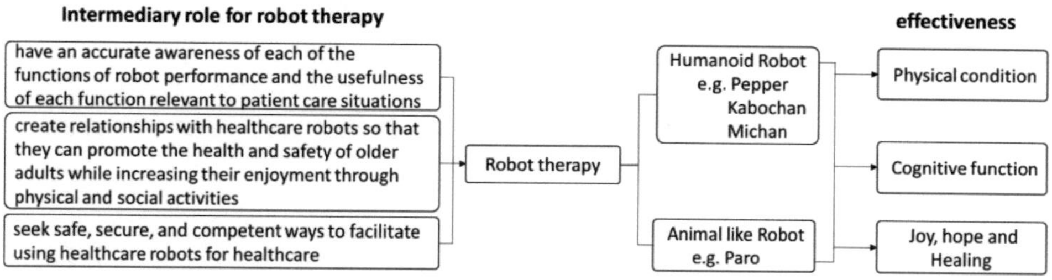

Figure 2.
The framework for robot therapy program.

By interaction with humanoid or animal-like robot, the older adults who are not physically active may have the improvement of their physical condition [31], such as hugging and stroking them, talking with them, and participating in any activity involving the robot.

The intermediary role of the nurse involves mediating and connecting patients with robots. It also involves a focus on ethical and moral issues inherent in nursing situations that include activities by healthcare robots [25]. Specifically, the intermediary person was in charge of connecting the subject with the robot. An intermediary can support older persons to interact well with the robot according to their physical condition. Also, they can provide joy for older adults when interacting with the robot, and among other persons. Moreover, in this interaction, the cognitive function of older persons with dementia may improve as they are able to communicate or have a conversation with robots [32]. Thus, intermediary can ask older adults if they are having fun, and if they feel like they have a companion in their daily life.

In Japan, various robots are produced and introduced for robot therapy in hospitals and other health care facilities. However, the performance and functions of these robots are often of lesser fidelity and functionality than expected by some facilities, thereby preventing their continued use after the initial introduction [33, 34]. The dialogue between healthcare robots and older adults was difficult without an intermediary role because of the difficulty for older adults to understand the robot because of the speed of its speech and tone of vocalization [15]. Oftentimes, because of this robot inefficiency, it is essential to consider instituting the intermediary role of nurses to engage the robot with the older adults. This nurses' role can enhance utility and instigate efficiency even if the robot is not sophisticated enough to be useful for rehabilitation and dialogue with older adults.

4. Method

4.1 Subjects of the study

The subjects of the study were two female older persons who were diagnosed with dementia using the Hasegawa's Dementia Scale-Revised (HDS-R) [35] instrument. Both were in their 80s and met the following inclusion criterion (diagnosed with dementia with a certain score): the HDS-R between the score range of 3-20 points. Exclusion criteria included older people who could not communicate verbally, those who could not interact with dogs and small stuffed toy robots, those who could not wear a portable electrocardiogram, and those who could not consent from their families.

- Subject A received the animal therapy intervention. She was in her 80's and diagnosed with dementia with a HDS-R score of 8 points.

- Subject B received the robot therapy. She was in her 80's and diagnosed with dementia with a HDS-R score of 10 points.

Data collection occurred on a single day, in a single observation period for each person; data collection for animal therapy was on October 10, 2017, and for robot therapy was October 25, 2019.

4.2 Hypothesis

As a prediction of the data comparison results, both animal therapy and robot therapy have the same effect on the physical, mental, and cognitive functions of the older person. Each subject data was extracted from animal and robot therapy, and both effects were compared. Data extraction methods were indicated in **Figures 3** and **4**.

4.3 Ethical consideration

The data collection procedure was performed following the Private Information Protection Law, with approval from the Tokushima University Hospital Ethics Board (approval number 2039) and Mifune Hospital (approval number 20170201-1). The purpose and methods used in the study were explained to all subjects and their guardians. Subjects were assured that their personal information would be protected and would only be used for research purposes, and that anonymity would be maintained in the report.

4.4 Data extraction method

4.4.1 Animal therapy protocol

As described below, a total of 5 minutes of data was extracted before, during, and after the therapy (**Figure 3**). In the animal therapy, the subject was an older person with dementia who was admitted to the facility. Animal therapy was performed by a therapist after music therapy. The animal used was a dog.

4.4.2 Robot therapy protocol

As described below, robot therapy was set to a total of 5 minutes of data before, during, and after the therapy. However, due to data collection constraints, it could not obtain data for 5 minutes during and after the therapy (**Figure 4**). In the robot therapy procedure, the subject was an older person with dementia. The robots used were Kabo-chan (W23×H28×H28 (sitting high), weight 680g) and Mi-chan (W25×D20×H 30 (sitting high) 30cm, weight 390g). These robots can talk, sing, and slightly nod charmingly in response to touch and spoken words. These robots have sensors that are installed in the mouth, head, hands, feet, and main body. These sensors allow the robots to verbally respond to any sounds and movements.

4.5 Procedure of data analysis

4.5.1 Analysis of autonomic nervous activity

Heart Rate Variability (HRV) data were assessed at various frequency bands using an HRV software tool (MemCalc/Bonaly Light: GMS, Tokyo, Japan).

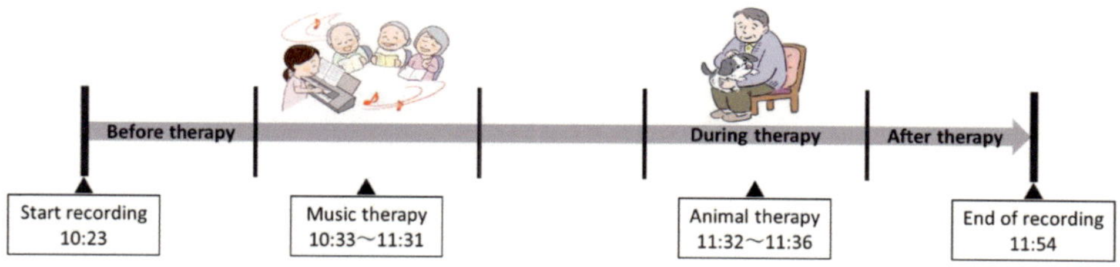

Figure 3.
Animal therapy protocol. Animal therapy data extraction method. Before therapy 10:24:00 ~ 10:28:58 (5 minutes in total); During therapy 11:32:00 ~ 11:36:58 (5 minutes in total); After therapy 11:37:00 ~ 11:41:58 (5 minutes in total). As described above, a total of 5 minutes of data was extracted before, during, and after the therapy.

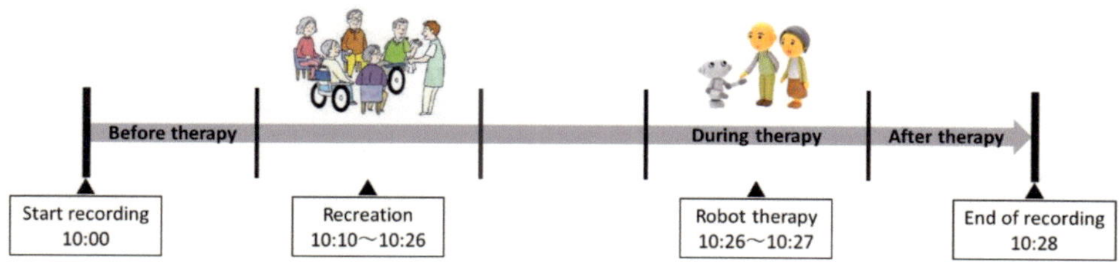

Figure 4.
Robot therapy protocol. Robot therapy data extraction method. Before therapy 10:05:00 ~ 10:09:58 (5 minutes in total); During therapy 10:26:16 ~ 10:27:58 (1 minute 41 seconds in total); After therapy 10:27:58 ~ 10:28:40 (42 seconds in total). As described above, it was set to extract a total of 5 minutes of data before, during, and after the therapy. However, due to data constraints, data during and after the therapy was less than 5 minutes. The measurement time was sometimes short because the measurement was performed according to the procedure of robot therapy in the clinical setting. In addition, an artifact was included in the electrocardiogram due to the subject's movements, so it was not possible to obtain all the data for 5 minutes. This was the limitation of this clinical study.

The low frequency (LF) and high frequency (HF) bands in heart rate variability (HRV) reflect sympathetic and parasympathetic nervous systems which is commonly accepted as the activities of the autonomic nervous system [36, 37]. In a continuously recorded data, inter-beat (R-R) intervals were obtained for a 1-min segment using the maximum entropy method. In this study, the two major spectral components of HRV, the variances of the Low-Frequency (LF: 0.04 - 0.15 Hz) band and High-Frequency (HF: 0.15 - 0.4 Hz) band, were calculated. The HF data can be used as an index of parasympathetic nervous activity, and the LF/HF ratio can be used as an index of sympathetic nervous activity.

An optimal level of variability within an organism's key regulatory systems is critical to the inherent flexibility and adaptability or resilience that epitomizes healthy functioning and well-being [38]. HRV is the change in the time intervals between adjacent heartbeats. It is an emergent property of interdependent regulatory systems that operate on different time scales to adapt to environmental and psychological challenges. The heart's rhythms are characterized as reflecting both physiological and psychological functional status of internal self-regulatory systems. Lowered parasympathetic activity, rather than reduced sympathetic functioning appears to account for the reduced HRV in aging [39]. This can be observed when persons engage in meeting a challenge that requires effort and increased sympathetic activation. Alternatively, it can indicate increased parasympathetic activity as occurs during slow breathing [40]. With psychological regulation, lower HF power is associated with stress, panic, anxiety, or worry [41].

In this study, the use of HRV was critical in measuring the psychological functional status and emotional experience of older persons particularly those

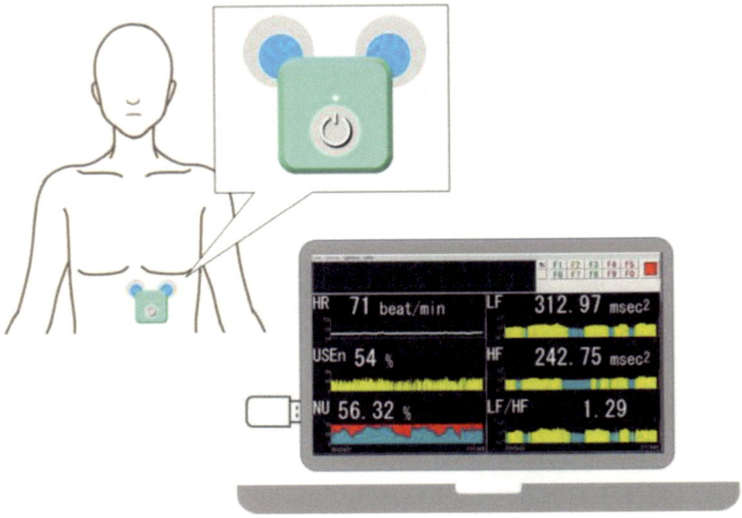

Figure 5.
Heart Rate Variability (HRV) data were assessed at various frequency bands using an HRV software tool (MemCalc/Bonaly Light: GMS, Tokyo, Japan).

with dementia. Changes in autonomic nervous activity were determined using HRV. Heart Rate (HR) -mean, HF, LF/HF, NU % (ratio of sympathetic nerve components), and body movement (acceleration) are shown in the graph, it shows the relationship between subjects' interaction with dog or robot doll during intervention. The results were recorded graphically, enabling visual assessments and measurements (**Figure 5**).

5. Typical examples and discussions

5.1 Comparison of animal therapy and robot therapy using the HRV and accelerometer as biological responses of older persons with dementia

5.1.1 Animal therapy

5.1.1.1 The experimental data before the animal therapy

The data were visually recorded enabling graphic visual assessments and measurements. The analysis of the HRV allowed the evaluation of autonomic nervous function.

After the movement of the dog (animal) trunk, the sympathetic nerve recordings became predominant. Afterwards, it was the parasympathetic nerve that became predominant. Changes in autonomic nervous activity can be confirmed from the data recorded in the accelerometer and the results of heart rate variability analysis as the body moves (**Figure 6**).

5.1.1.2 The experimental data during the animal therapy

Subject A's HRV data showed sympathetic nerve predominance after touching the dog, after which it showed a sympathetic nerve predominance. Before touching the dog for the second time, the pulse rate decreased, and the parasympathetic nerve became predominant. By interacting with the dog, fluctuations in heart rate were observed,

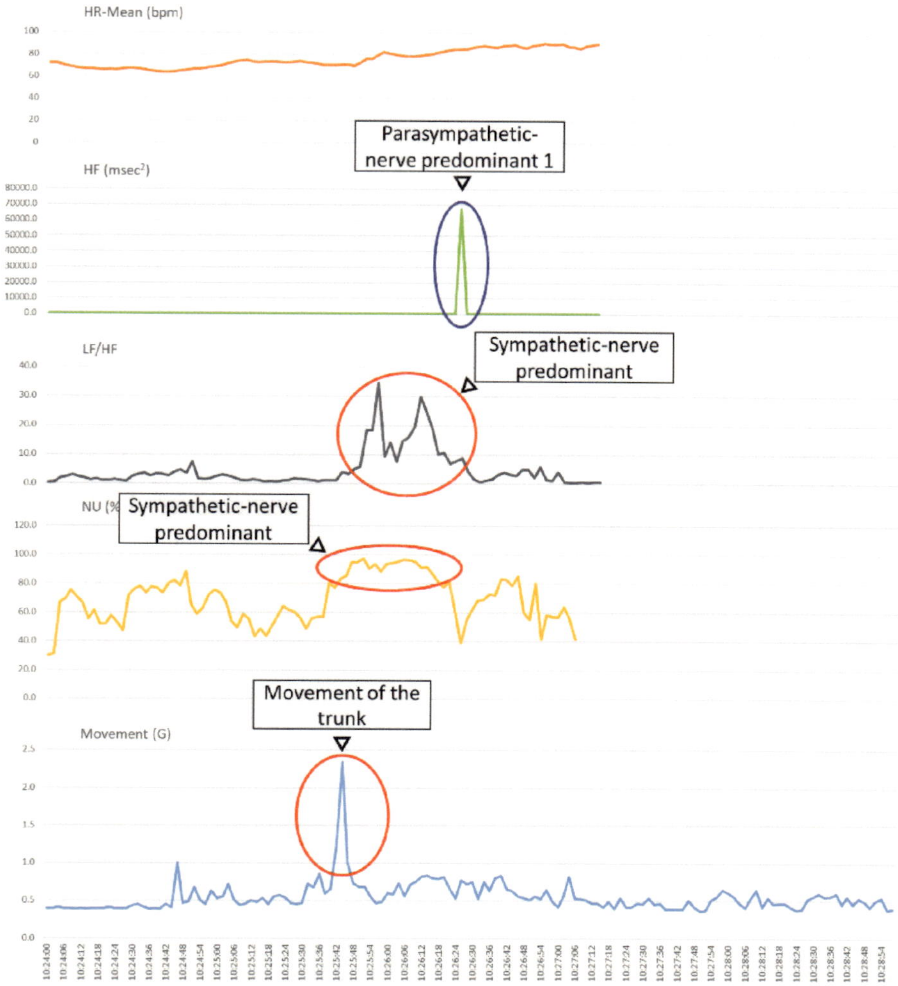

Figure 6.
The experimental data before the animal therapy (Subject A). Note: HR-mean; Heart Rate mean, HF; High Frequency, LF/HF; Low Frequency / High Frequency, NU; Normalized Unit, Movement; the body movement. The unit of vertical axis. HR-Mean; Beat/min, HF; msec2, LF/HF; Nothing, NU (%), Movement; G. The unit of horizontal axis; hour: min: second.

and a balance between sympathetic and parasympathetic nerve activities was observed. Therefore, it was considered that effective stimulation could be provided by contact between the subject and subject A through interaction with the dog (**Figure** 7).

5.1.2 Robot therapy Data

5.1.2.1 The experimental data before robot therapy

Figure 8 shows the alternating sympathetic and parasympathetic activities before the robot therapy.

5.1.2.2 The experimental data during the robot therapy

After holding the robot, it was observed that the parasympathetic nerve activity became predominant, and then subsequently, the sympathetic nerve became predominant. In addition, after holding the robot for the second time, the parasympathetic nerve became predominant.

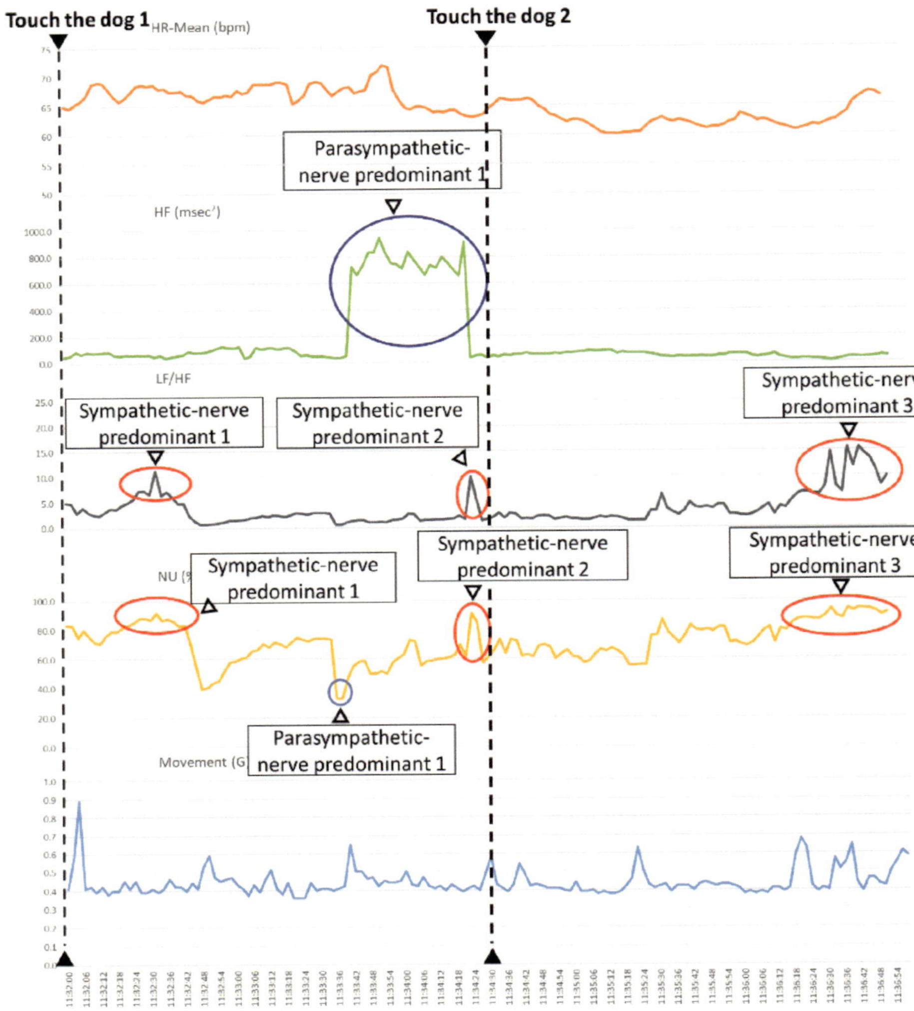

Figure 7.
The experimental data during animal therapy (Subject A).

Sympathetic nerve activity was predominant before the start of the therapy. However, during the therapy, the sympathetic nerve activity predominance continued immediately after holding the robot was observed. Autonomic nervous activity was stable at the end of the therapy (**Figure 9**).

5.1.2.3 Comparison of the interactions between older people with dementia and their caregivers during animal therapy and during robot therapy

In both animal therapy and robot therapy, stable heart rate and body movements were confirmed in all processes before, during, and after therapy. These were during the awake state, and the awakening could be confirmed visually by participant observation and recorded video data.

In the animal therapy, the LF/HF value was high even before the start of therapy, and the predominance of sympathetic nerve activity was confirmed. During the therapy, the LF/HF value increased immediately after the first touch of the dog, immediately before the second touch of the dog, and immediately after the touch. These activities confirmed that the sympathetic nerve activities were dominant.

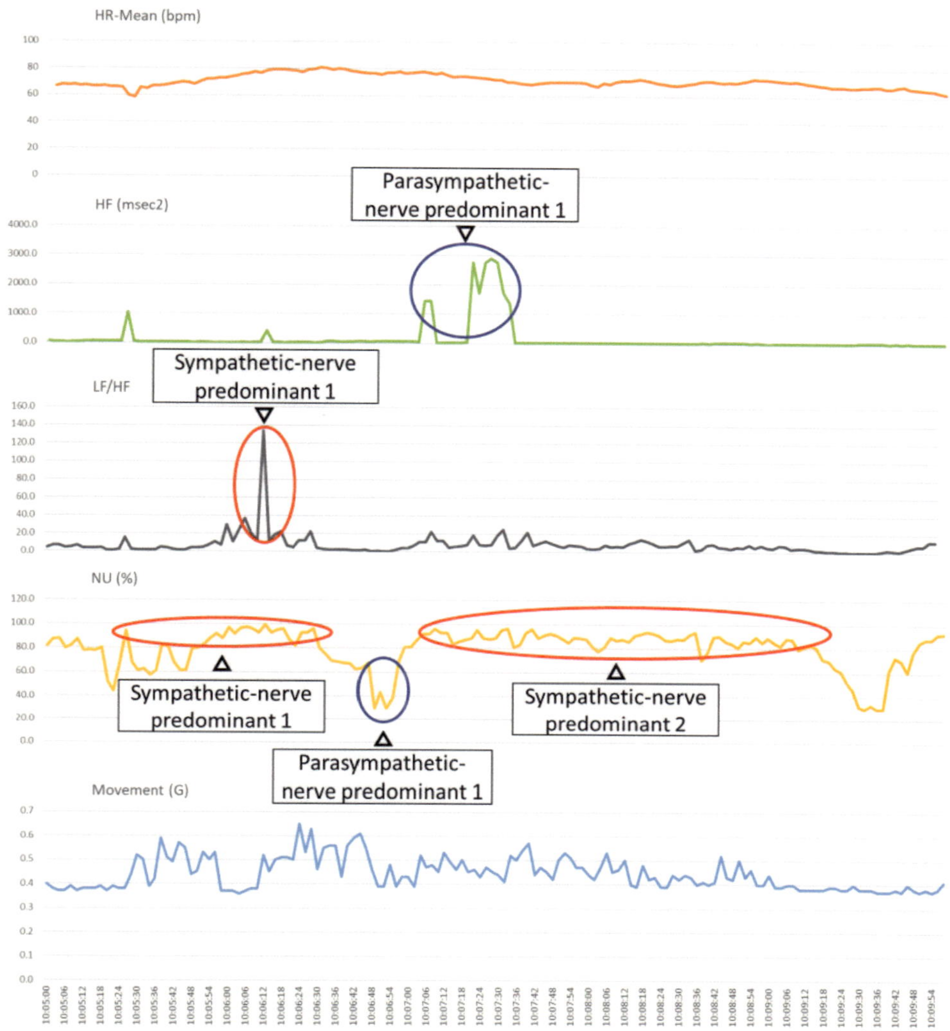

Figure 8.
The experimental data before robot therapy (Subject B).

At the end of the therapy, the LF/HF value was high for about one minute, confirming the predominance of sympathetic nerve activity.

In the robot therapy, the subjects had high LF/HF values and predominant sympathetic nerve activity before the start of the therapy. In addition, during the therapy, high LF/HF values that continued immediately after the robot was first held were observed, confirming the predominance of sympathetic nerve activity. After the therapy, the autonomic nervous activity became stable.

5.2 Comparison intermediary interactions during animal therapy and robotic therapy using participant observations with older adults

Animal therapy was conducted by the therapist and pianist in the healthcare institution. After music therapy, the therapist brought a dog to the older person (Subject A). Her expression can be seen from **Figure 10**. From the observation, it was evident that the older person spontaneously touched and stroked the dog. Subject A seemed happy touching the dog and intermediary bring the dog near her.

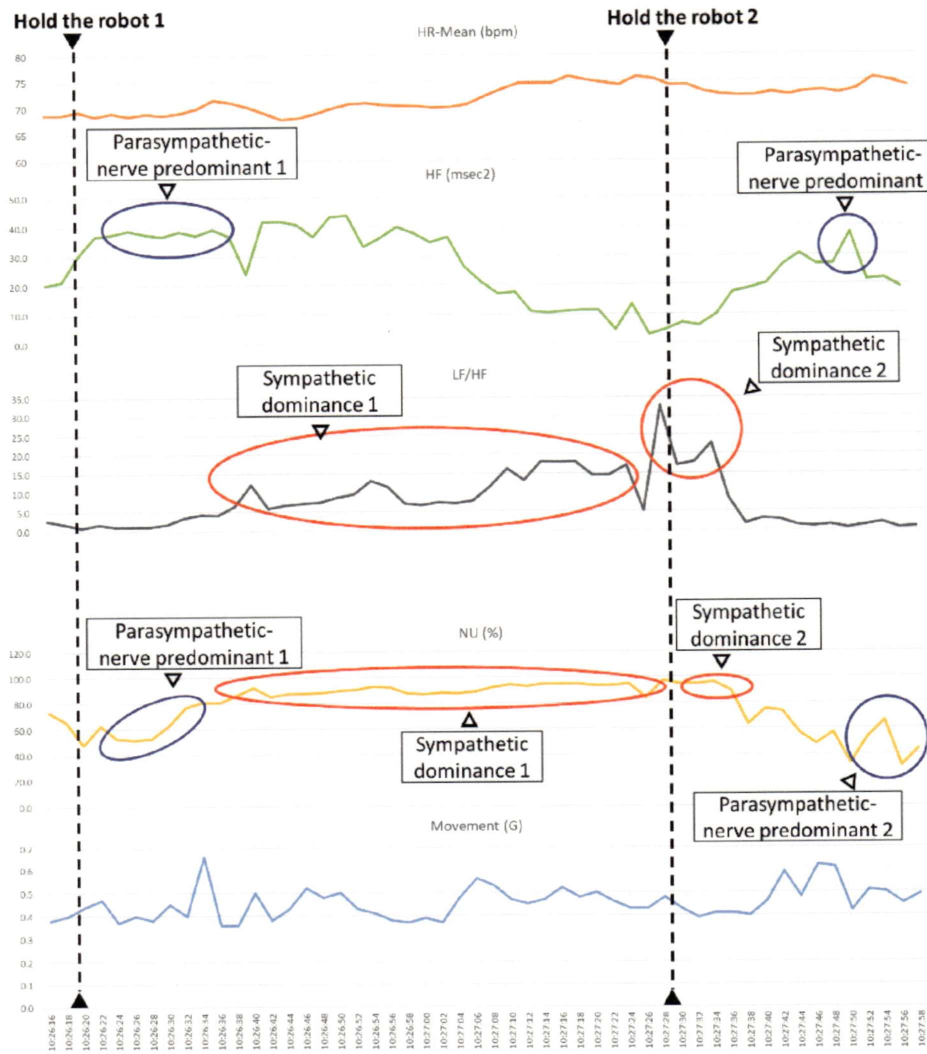

Figure 9.
The experimental data during robot therapy (Subject B).

The nurse intermediary was in charge of connecting Subject A with the dog. The intermediary asked Subject A some questions: "Do you like dogs and animals" and "Have you ever owned a dog?" Subsequently, the intermediary asked, "May I bring the dog closer to you?" Since Subject A's mobility requires a cane when walking, and her daily activities is slow due to old age, the intermediary held on to the small dog, and brought it near to subject A's chest (**Figure 10**) to make it easier for her to touch the dog. Subject A replied that she had a dog, and that she liked dogs and had many dogs, and that one of her dogs was as small as the size of dog used for animal therapy. Then the intermediary picked a conversational topic, and the dog waved its tail seemingly to mean that "This dog seems happy."

Subject B was in the robot therapy section. It was observed that older persons seemed happier during their interactions with robots. Subject B touched and held the robots. She stroked their legs and arms, and head as if the robots were her grandchildren. When the older person saw the pictures displayed in the television screen, they turned the robot to see the TV screen and she exclaimed to the robot "Look at the TV!" After that, she asked the robot some questions like "Do you like animals?" And she stroked the robot's head. When the robot said, "Thank you", she

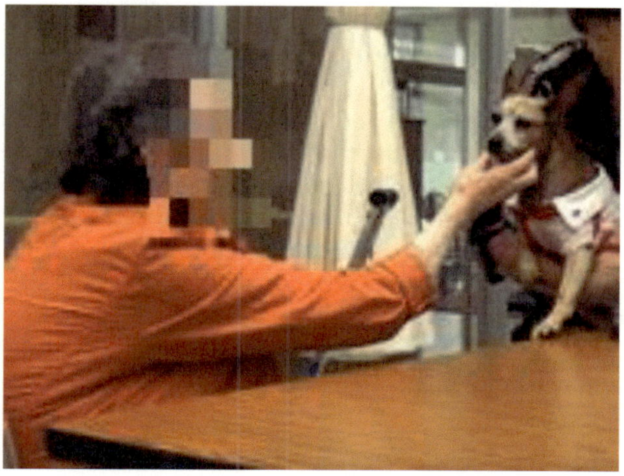

Figure 10.
Subject A's scene during the animal-assisted therapy. During animal therapy: she spontaneously stretched her arms, stroked dog's body, and smiled from beginning to end.

laughed. When the robot sang, she clapped her hands and said, "You're so good!" (**Figure 11**).

The intermediary asked subject B about her impression of robots and if she was interested in the robot. Subject B was interested in the robot and listened to what the robot had to say. She replied, "around 80 years old" to the robot question, "How old are you?" However, there were some occasions when she could not hear what the robot said. Therefore, the intermediary had to repeat what the robot said. Since the inexpensive robot used had limited conversational word content, the intermediary supplemented the conversation content and enhanced the interval in the conversation. Also, since the robot can sing songs, the intermediary tried to sing thereby illustrating that it was the robot that was singing. Subject B listened intently while the robot was singing, and she enjoyed singing to the tune along with the robot. On one occasion, the intermediary informed subject B that she could touch the robot. She asked. "Would you like to pick up robot Mi chan?"

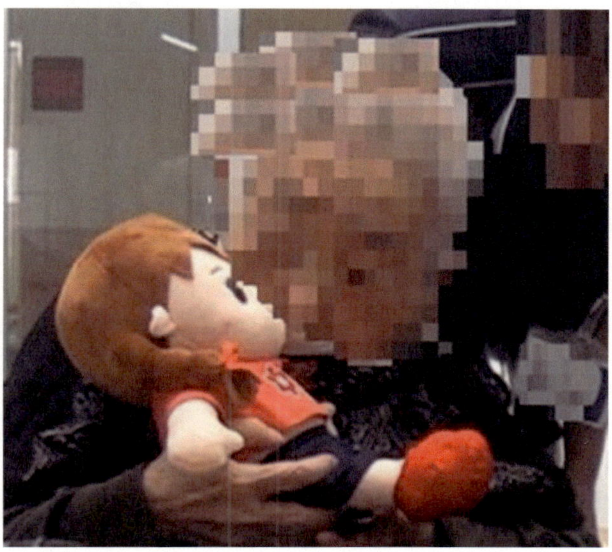

Figure 11.
Subject B′ scene during robot therapy. During robot therapy: she hugged robot, smile and talked to the robot.

Conversations with the robots illustrated that nurses as intermediaries can show that emotional conversations establish effective transactive engagements between subjects and robots.

The comparison of AAI and Robot therapy showed that each method has its benefits and shortcomings indicating that the two methods could potentially complement each other. Both therapies were shown to have a possibility of beneficial effects on the emotional wellbeing of patients with dementia. There is a possibility that if robot therapy using an inexpensive robot such as used in this study might be obtain the same effect as AAI, barriers peculiar to AAI such as zoonotic diseases, animal bites, and allergies can be avoided. In addition, it will be possible to use AAI and robot therapy properly while taking advantage of their respective characteristics and advantages.

6. Conclusion

This chapter explained the robot therapy program for patients with dementia from the viewpoint of its framework and effectiveness.

The electrographic data provided neurophysiological evidence of the influence of robot utilization on the autonomic nervous system activity of older adults with dementia. The examples described were demonstrations of studies, which captured how data were collected through different devices and specific procedures to describe, explain, predict, and prescribe phenomena, as evidenced from a rigorous analysis of data regarding human-robot interaction with nurses as intermediaries.

The typical examples show that animal therapy has almost the same effectiveness as robot therapy among older people. It is clarified that robot therapy can be expected to have a healing effect on patients, improve motivation for activity, and increase the amount of activity, similar to animal therapy. Furthermore, it was essential to consider the intermediary role of nurses for connecting the robot and older adults and their role, even if the robot is not sophisticated enough to be useful as a humanoid nurse robot for rehabilitation and dialogue with older adults.

Thus, robot therapy could be considered another important intervention in the challenging health and innovative care practices needed in the care of older persons. Nevertheless, two issues were realized regarding living with a human-type communication robot as a strategy for rehabilitation care and to improve cognitive functions and prevent cognitive decline in the older adults. In this regard, robot therapy has not been generalized, and more analysis, descriptions and discussions about its practical utility are required.

Acknowledgements

This work was supported by JSPS KAKENHI Grant Numbers JP17K17504. We would like to express our sincere appreciation to all the patients and participants who contributed to this article. With many thanks to Dr. Savina Schoenhofer for reviewing this chapter prior to its publication.

Author details

Kyoko Osaka[1*], Ryuichi Tanioka[2], Feni Betriana[2], Tetsuya Tanioka[3], Yoshihiro Kai[4] and Rozzano C. Locsin[3]

1 Department of Clinical Nursing, Kochi Medical School, Kochi University, Kochi, Japan

2 Graduate School of Health Sciences, Tokushima University, Tokushima, Japan

3 Department of Nursing, Institute of Biomedical Sciences, Tokushima University Graduate School, Tokushima, Japan

4 Department of Mechanical Engineering, Tokai University, Kanagawa, Japan

*Address all correspondence to: osaka@kochi-u.ac.jp

References

[1] Muramatsu N, Akiyama H: Super-aging society preparing for the future. The Gerontologist. 2011;51(4):425-432. DOI: 10.1093/geront/gnr067

[2] Ministry of Health, Labour, and Welfare. 2015 Edition, Annual Health, Labour and Welfare Report-Consideration of a depopulating society- Towards a society where people can live in peace and realize of their hope. 2015. https://www.mhlw.go.jp/english/wp/wp-hw9/dl/summary.pdf [Accessed: 2020-12-25]

[3] Department of International Affairs Japanese Nursing Association: Nursing in Japan 2016. https://www.nurse.or.jp/jna/english/pdf/nursing-in-japan2016.pdf

[4] Buchan J, Aiken L: Solving nursing shortages: A common priority. J Clin Nurs. 2008;17(24): 3262-3268. DOI: 10.1111/j.1365-2702.2008.02636.x [Accessed: 2020-12-25]

[5] Amjad H, Carmichael D, Austin AM, Chang C, Bynum JPW: Continuity of care and health care utilization in older adults with dementia in Fee-for-Service Medicare. JAMA Intern Med. 2016;176(9):1371-1378. DOI: 10.1001/jamainternmed.2016.3553

[6] Mangalavite AM: Animal-Assisted Therapy: Benefits and Implications for Professionals in the Field of Rehabilitation. 2014;1-26. Research Papers. Paper 547. http://opensiuc.lib.siu.edu/gs_rp/547

[7] Paula Calvo, Joan R. Fortuny, Sergio Guzmán, Cristina Macías, Jonathan Bowen, María L. García, Olivia Orejas, Ferran Molins, Asta Tvarijonaviciute, José J. Cerón, Antoni Bulbena, Jaume Fatjó: Animal Assisted Therapy (AAT) Program As a Useful Adjunct to Conventional Psychosocial Rehabilitation for Patients with Schizophrenia: Results of a Small-scale Randomized Controlled Trial. Frontiers in Psychology. 2016;7:1-13. DOI: 10.3389/fpsyg.2016.00631

[8] Thodberg K, Sørensen L.U, Christensen J.W, Poulsen P.H, Houbak B, Damgaard V, Keseler I, Edwards D, Videbech P.B: Therapeutic effects of dog visits in nursing homes for the elderly. Psychogeriatrics. 2016;16(5):289-297. DOI: 10.1111/psyg.12159

[9] Richeson NE: Effects of animal-assisted therapy on agitated behaviors and social interactions of older adults with dementia. Am J Alzheim Dis Other Dement. 2003;18(6):353-358. DOI: 10.1177/153331750301800610

[10] Park S, Bak A, Kim S, Nam Y, Kim H, Yoo DH., Moon M: Animal-assisted and pet-robot interventions for ameliorating behavioral and psychological symptoms of dementia: a systematic review and meta-analysis. Biomedicines. 2020;8(6):150. DOI: 10.3390/biomedicines8060150

[11] Minmin L, Peng L, Ping Z, Mingyue H, Haiyan Z, Guichen L, Huiru Y, Li C: Pet robot intervention for people with dementia: a systematic review and meta-analysis of randomized controlled trials. Psychiatric Research. 2019;271:516-525. DOI: 10.1016/j.psychres.2018.12.032

[12] Feast AR, White N, Candy B, Kupeli N, Sampson EL: The effectiveness of interventions to improve the care and management of people with dementia in general hospitals: a systematic review. Int J Geriatr Psychiatry. 2020;35(5):463-488. DOI: 10.1002/gps.5280

[13] Yokoyama A: Challenges for the future of robot assisted therapy from the viewpoint of psychiatry: with the knowledge of animal assisted therapy.

Journal of The Society of Instrument and Control Engineers. 2012 July 10; 51(7): 598-602.

[14] Osaka K, Sugimoto H, Tanioka T, Yasuhara Y, Locsin R, Zhao YR, Okuda K, Saito K: Characteristics of a transactive phenomenon in relationships among older adults with dementia, nurses as intermediaries, and communication robot. Intelligent Control and Automation. 2017; 8: 111-125. DOI: 10.4236/ica.2017.82009

[15] Osaka K, Tanioka T, Tanioka R, Kai Y, Locsin RC. Effectiveness of care robots, and the intermediaries' role between and among care robots and older adults. In: Proceedings of the IEEE/SICE International Symposium on System Integration (SII '20), 12-15 January 2020; January 2019; Honolulu, Hawaii: IEEE; 2020. p.611-616 DOI: 10.1109/SII46433.2020.9026262

[16] Tanioka R, Yasuhara Y, Osaka K, et al: Autonomic nervous activity of patient with schizophrenia during Pepper CPGE-led upper limb range of motion exercises. Enferm Clin. 2020;1:48-53. DOI: 10.1016/j.enfcli.2019.09.023

[17] American Veterinary Medical Association. Animal-assisted interventions: Definitions. (n.d.). Retrieved from https://www.avma.org/resources-tools/avma-policies/animal-assisted-interventions-definitions [Accessed: 2020-12-25]

[18] Nimer J, Lundahl B: Animal-assisted therapy: a meta-analysis. Anthrozoos. 2007;20(3): 225-238. DOI: 10.2752/089279307X224773.

[19] S H Hosseini, K M Goher: Personal Care Robots for Older Adults: An Overview. 2017;13:11-19. DOI: 10.5539/ass.v13n1p11

[20] Fiske A, Henningsen P, Buyx A: Your robot therapist will see you now: ethical implications of embodied artificial intelligence in psychiatry, psychology, and psychotherapy. J Med Internet Res. 2019;21(5). DOI: 10.2196/13216.

[21] Shibata T, Wada K: Robot therapy: a new approach for mental healthcare of the elderly - a mini review. Gerontology. 2011;57: 378-386. DOI: 10.1159/000319015

[22] Huei-Chuan Sung, Shu-Min Chang, Mau-Yu Chin, Wen-Li Lee: Robot-assisted therapy for improving social interactions and activity participation among institutionalized older adults: a pilot study. Asia Pac Psychiatry. 2015;7:1-6. DOI: 10.1111/appy.12131

[23] Birks M, Bodak M, Barlas J, Harwood J, Pether M: Robotic seals as therapeutic tools in an aged care facility: a qualitative study. J Aging Res. 2016;article ID 8569602:7 pages. DOI: 10.1155/2016/8569602

[24] Osaka K: Development of the model for the Intermediary role of nurses in transactive relationships with healthcare robots. Int J Hum Caring. 2020;24(4):265-274. DOI: 10.20467/HumanCaring-D-20-00014

[25] Yasuhara Y, Tanioka R, Tanioka T, Ito H, Tsujigami Y: Ethico-legal issues with humanoid caring robots and older adults in Japan. Int J Hum Caring. 2019;23(2):141-148. DOI: 10.20467/1091-5710.23.2.141

[26] Locsin R: Technological competency as caring in nursing. Sigma Theta Tau International Press. 2005. 229 p.

[27] Boykin A, Schoenhofer S: Nursing as caring: A model for transforming practice. 2nd ed. Jones and Bartlett. 2001. 71 p.

[28] Cherniack EP, Cherniack AR: The benefit of pets and animal-assisted therapy to the health of older individuals. Curr Gerontol

Geriatr Res. 2014;2014:623203. DOI: 10.1155/2014/623203

[29] Lundqvist M, Carlsson P, Sjödahl R, Theodorsson E, Levin LÅ: Patient benefit of dog-assisted interventions in health care: a systematic review. BMC Complement Altern Med. 2017;17(1):358. DOI:10.1186/s12906-017-1844-7

[30] K. Wada, T. Shibata, T. Saito, K. Tanie. Effects of robot-assisted activity for elderly people and nurses at a day service center. In: Proceedings of the IEEE, 2004;92(11):1780-1788, DOI: 10.1109/JPROC.2004.835378

[31] Fasola J, Mataric MJ: Using socially assistive human–robot interaction to motivate physical exercise for older adults. In: Proceedings of the IEEE, 2012;100(8):2512-2526, Aug. 2012, DOI: 10.1109/JPROC.2012.2200539

[32] Tanaka M, Ishii A, Yamano E, Ogikubo H, Okazaki M, Kamimura K, Konishi Y, Emoto S, Watanabe Y: Effect of a human-type communication robot on cognitive function in elderly women living alone. Med Sci Monit. 2012;18(9):CR550-557. DOI: 10.12659/msm.883350

[33] Tanioka T, Smith CM, Osaka K, Zhao Y: Framing the development of humanoid healthcare robots in caring science. Int J Hum Caring. 2019;23(2):112-120, 2019. DOI: 10.20467/1091-5710.23.2.112

[34] Ferreira FMRM, Chaves MEA, Oliveira VC, Van Petten AMVN, Vimieiro CBS: Effectiveness of robot therapy on body function and structure in people with limited upper limb function: a systematic review and meta-analysis. PLoS One. 2018;13(7):e0200330. DOI: 10.1371/journal.pone.0200330.

[35] Imai Y, Hasegawa K: The revised Hasegawa's Dementia Scale (HDS-R)-Evaluation of its usefulness as a screening test for dementia. Hong Kong Journal of Psychiatry. 1994;4(SP2):20-24.

[36] von Rosenberg W, Chanwimalueang T, Adjei T, Jaffer U, Goverdovsky V, Mandic DP: Resolving ambiguities in the LF/HF ratio: LF-HF scatter plots for the categorization of mental and physical stress from HRV. Front in Physiol. 2017;.8:360. DOI: 10.3389/fphys.2017.00360

[37] Takahashi N, Kuriyama A, et.al: Validity of spectral analysis based on heart rate variability from 1-minute or less ECG recordings. Electrophysiology. 2017;40:1004-1009. DOI: 10.1111/pace.13138

[38] Fred Shaffer, Rollin McCraty, Christopher L. Zerr: A healthy heart is not a metronome: an integrative review of the heart's anatomy and heart rate variability. frontiers in psychology. 2014;5:1-19. DOI: 10.3389/fpsyg.2014.01040

[39] Umetani K, Singer DH, McCraty R, Atkinson M. Twenty-four hour time domain heart rate variability and heart rate: relations to age and gender over nine decades. J Am Coll Cardiol. 1998;31(3):593-601. DOI: 10.1016/s0735-1097(97)00554-8

[40] Marc A. Russo, Danielle M. Santarelli, Dean O'Rourke: The physiological effects of slow breathing in the healthy human. Breathe. 2017;13(4): 298-309. DOI: 10.1183/20734735.009817

[41] Rollin McCraty, Fred Shaffer. Heart Rate Variability: New Perspectives on Physiological Mechanisms, Assessment of Self-regulatory Capacity, and Health risk: Glob Adv Health Med. 2015;4(1): 46-61. DOI: 10.7453/gahmj.2014.073

Section 4

Building Artificial Intelligence in Robots and the Ethics of Healthcare Robots

Chapter 7

Artificial Brain for the Humanoid-Nurse Robots of the Future: Integrating PsyNACS© and Artificial Intelligence

Hirokazu Ito, Tetsuya Tanioka, Michael Joseph S. Diño, Irvin L. Ong and Rozzano C. Locsin

Abstract

Robots in healthcare are being developed rapidly, as they offer wide-ranging medical applications and care solutions. However, it is quite challenging to develop high-quality, patient-centered, communication-efficient robots. This can be attributed to a multitude of barriers such as technology maturity, diverse healthcare practices, and humanizing innovations. In order to engineer an ideal Humanoid-Nurse Robots (HNRs), a profound integration of artificial intelligence (AI) and information system like nursing assessment databases for a better nursing care delivery model is required. As a specialized nursing database in psychiatric hospitals, the Psychiatric Nursing Assessment Classification System and Care Planning System (PsyNACS©) has been developed by Ito et al., to augment quality and safe nursing care delivery of psychiatric health services. This chapter describes the nursing landscape in Japan, PsyNACS© as a specialized nursing database, the HNRs of the future, and the future artificial brain for HNRs linking PsyNACS© with AI through deep learning and Natural Language Processing (NLP).

Keywords: PsyNACS©, artificial brain, humanoid-nurse robot, artificial intelligence, communication, nursing

1. Introduction

The nursing shortage in Japan has significantly increased in years [1]. Population aging coupled with a declining birthrate has greatly steered the upward demand for nursing professionals [2, 3]. In response, the Government of Japan has called for the adoption of the Internet of Things (IoT) and robots in healthcare [4, 5]. Japan's Act for the Mental Health and Welfare of Persons with Mental Disorders has been undertaken reducing chronic psychiatric hospital stay as much as possible and providing home care services. However, the nursing shortage has created can be more pressing healthcare issues in psychiatric hospitals where the length of average hospital stay is much longer in contrast with other countries [6, 7].

Psychiatric signs and symptoms provide dysfunctional evidence and pose as healthcare disadvantage to patients with mental illness [8–10]. These dysfunctions and disadvantages characterized by repeated exacerbations and remissions for patients with mental illness often follow a chronic course [11] wherein they maintain their lives in the community while being repeatedly admitted to and discharged from psychiatric hospitals [12, 13]. In addition, psychiatric nursing care situations have become more challenging as the number of people with dementia continues to increase with the population getting older.

Caring for patients with dementia is complex and requires specialized interventions [14]. In Japan, these situations influence psychiatric healthcare services therefore, psychiatric nursing practice requires early assessment during the acute phase, and effective health maintenance in the chronic phase to provide optimal nursing care for patients.

Access to health data is essential in clinical decision making. This can considerably improve interdisciplinary care and health outcomes of patients with psychiatric conditions, preventing unnecessary readmissions and unsafe discharges in psychiatric hospitals. In the absence of a functioning database, psychiatric nursing care becomes inefficient and fragmented, further creating an inadequate environment for IoT and robots in healthcare to thrive. On the whole, the current nursing landscape remains to exhibit countless unstructured challenges in attaining and sustaining quality psychiatric nursing care.

2. The PsyNACS© database

The need for a specialized nursing database for psychiatric hospitals in Japan prompted the development of the Psychiatric Nursing Assessment Classification System and Care planning System (PsyNACS©) to improve psychiatric nursing care services (Ito et al.) [15, 16]. This was a data-driven classification system of nursing assessment data for Japanese psychiatric healthcare which can be used in various patient care situations in psychiatric units.

In developing the system, a select group of experienced nurses (N = 664) working in psychiatric hospitals evaluated 211 assessment items for psychiatric nursing care derived from contemporary nursing theoretical models and frameworks. The results of the factor analysis of the final 209 assessment items generated 9 Patient Assessment Data (PAD) and 31 Cluster Assessment Data (CAD). Each PAD consisted of 2 to 5 CADs.

The PADs are simple categories for each corresponding CADs. The PADs include (a) psychological symptom and stress, (b) information about treatment, (c) function of eating and balance of water, (d) life and value, (e) vital signs and health assessment, (f) self-care, (g) social support, (h) activity, sleeping and mobility capability, and (i) Sexual function and sexual behavior.

PsyNACS© is designed to assist nurses to provide timely, effective and appropriate care for patients with mental illness. It can be a server-type, laptop-type, or web-type system. A server-type PsyNACS© installs a server in a psychiatric hospital and the laptop-type PsyNACS© can be used without an internet connection. Of interest, the web-type PsyNACS© is connected to a cloud server that enables online nursing care planning. These pathways address access to design care plans to meet the needs of the patient with mental illnesses for individualized care, including treatment, rehabilitation, and post-discharge welfare services. Since the PsyNACS© database deals with big data, it has a secure mechanism to gather healthcare information and other assessment data.

As a result, the completed database (**Figure 1**) was digitized so that nursing care plans can be accessed using a computer or laptop system.

Figure 1.
PsyNACS© login page.

PysNACS© works at a lower cost as compared to hospital systems where expensive electronic medical records may not be feasible. The web-type PsyNACS© deals with aggregate patient information and health data. This calls for accurate recording and proper documentation of health information, psychiatric symptoms, and care plans. PsyNACS© is equipped with a database content filter that displays alerts for capturing inappropriate words, entries, or inputs.

Like with any medical and health records, information security and ethics remain to be the greatest concern. Both systems and risks management can be difficult tasks for nursing or any health professionals. In essence, only patients and their lawful representatives may request information. As a result, confidential personal information can be managed separately from the psychiatric nursing assessments and care plans to prevent unauthorized disclosure and leaks. PsyNACS© uses a dedicated internet line or a Virtual Private Network (VPN) to ensure information safety.

Handling patient information in healthcare settings requires additional levels of protection for privacy and confidentiality. A top the terminal management on the client end, a third-party risk management system such as antivirus software is highly recommended against data breaches. Other security features include establishing quality procedures for handling electronic information, setting identity and password management procedures for authorized access, clarifying server management procedures to oversee servers, and ensuring the security of communication paths.

PsyNACS© offers a holistic approach to psychiatric nursing assessment. It collects health information to have a better assessment of patient needs and determine the most appropriate nursing interventions. PsyNACS© is organized strategically into information blocks (**Figure 2**) for (a) common in psychiatry, (b) with dementia, (c) with complication, and (d) additional information. This is achieved by integrating the PADs as the *key areas* of psychiatric assessment (left column) and the CADs as the *subareas* with corresponding nursing assessment items (right column). Thus, patient-centered nursing care can be planned and delivered using the psychiatric assessment with 9 areas, 31 subareas, and 209 items.

The database prototype for PsyNACS© displays the recommended and relevant health information that has been entered recurrently into the system as individual-level data. This requires active participation and utilization among professional nurses at the point of care. By weighing critical information from psychiatric nurses, it becomes possible for nurses to use the system with greater usability and functionality. The aggregate data in the database will grow eventually into big data, which can be analyzed for quality improvement.

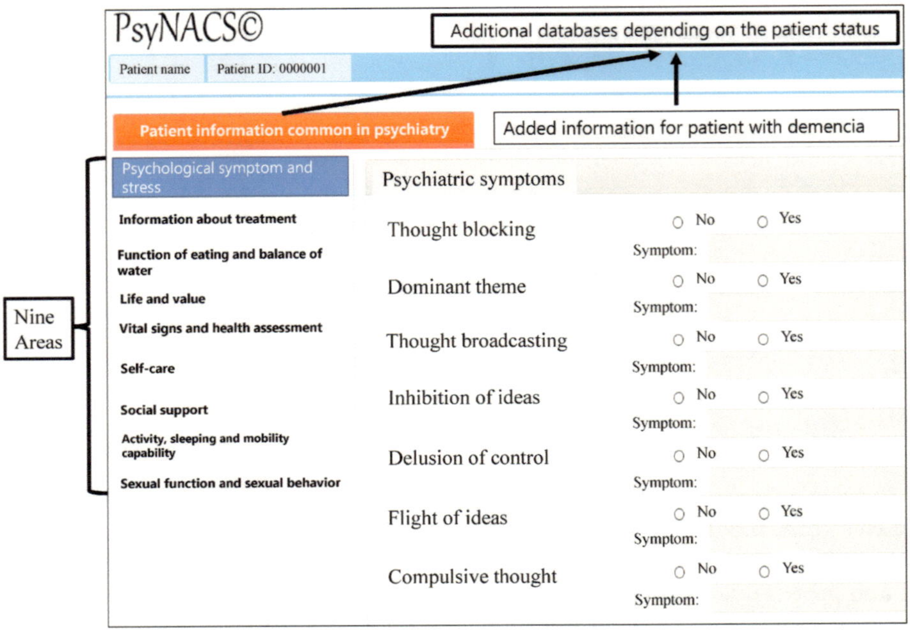

Figure 2.
PsyNACS© sample screen for area 1 - psychiatric symptom and stress.

In practice PsyNACS© was evaluated at a selected psychiatric hospital in Japan. Ten nurse managers who were experts in manipulating electronic medical charts in their respective psychiatric hospital participated and answered the questionnaire. They entered patient information data using the laptop-type of PsyNACS©. Evaluative processes included operability and efficiency of the system determined through the survey questionnaire. Five of the 10 participants responded that the system was good. Four participants declared that the information input method was efficient. However, regarding time required for inputting data was found to be significantly different among individual participants. Familiarity of the system operation was the main determinant (presented at the International Conference on Ethics, Esthetics, and Empirics in Nursing, Songkhla, Thailand, July, 5–7. 2017). Integrating PsyNACS© into nursing practice will provide nurses with better access to health information that allow them to perform holistic assessment and provide quality care that is responsive to current standards and contexts of Japanese psychiatric nursing.

3. Humanoid-nurse robots of the future

The Government of Japan's direction for robots in healthcare strongly coincides with the Fourth Industrial Revolution (4IR) [17] and Society 5.0 [18] in which a massive integration of highly advanced and recognized disruptive technologies such as AI, IoT, and quantum computing is expected to flourish. Nevertheless, this is quite challenging for healthcare, more so in psychiatric nursing which is lagging behind the manufacturing and other service industries. In order to thrive, information and communication technology are crucial for using humanoid robots in healthcare area. Despite technological advancements, the maturity of existing ideologies of Humanoid-Nurse Robots (HNRs) are yet a forthcoming consideration [19].

The HNRs of the future has no single definition, morphology (form), and physiology (function). Rather than a concrete conception, the HNRs of the future is considered a product of the collective visions of nursing and healthcare leaders as they reimagine the future of healthcare demands. In the context of psychiatric nursing and in older adult care in Japan, the HNRs are expected to assist and work with nurses in carrying out healthcare tasks and activities. Having a clear vision of HNRs as the supreme technological advancement in healthcare, the demands are for robots to be programmed in such a way that it can independently perform nursing-related technical skills, and simultaneously demonstrate value-added expressions of humanness such as respect, compassion, empathy, and caring [20]. Therefore, HNRs are envisioned to be of high-quality, expressing patient-centeredness, and efficient with communication.

First, being of high-quality means that HNRs are data-driven achieved by linking PsyNACS© with HNRs. Credible health information should be captured from meaningful nurse–patient interactions. The PsyNACS© as the conversation database and along with relevant data such as electronic medical records, history and physical examination, laboratory, and radiology results can be linked to AI enhancing HNRs to acquire reliable databases regarding patients with mental illness and dementia. In addition to PsyNACS© integration with HNRs, the quality of HNRs is frequently influenced by the global proliferation of robots in healthcare, especially in hospitals, communities, and in home settings. Producing high-quality HNRs also means having data-driven policies and guidance on shared nurse–robot practices. Therefore, it becomes essential for nurses to play leadership roles in the design, implementation, and evaluation of nurse-robot partnerships [21] and eventually transform this into standards of nursing practices.

Secondly, HNRs have patient-centered designs [22, 23]. The ethical standards of human nurses are primarily grounded on the value of caring [24]. This impacts the quality of nursing care. In psychiatric nursing, nurses need to address both the physical and psychosocial needs of patients particularly those with mental illness and dementia. This poses greater challenges to HNRs which are originally designed as provider-centric in order to improve the efficiency of healthcare workers.

Patient-centered designs can be accomplished by focusing on patient needs and therapeutic conversational contents of professional nurses. Given that HNRs can be both a technological tool and a care partner, this will also entail looking into the code of ethics for nurses in Japan (Japanese Nursing Association, 2003). The current provision mentions nothing about HNRs; only focusing on collaboration among nursing and healthcare personnel. Therefore, the nurse-robot partnership should carefully consider and meet the nursing code of ethics [25].

Lastly, HNRs are capable of efficient communication. Currently designed/developed robots usually engage in one-way communications – each time simply asking one-sided questions to attempt a dialog. Optimizing structured conversations are needed to elicit desired levels of engagement and participation. This can be achieved through the creation of a "Caring Dialogue Database" for HNRs to provide better information about the patients, and to share experiences of human-robot interactions. Moreover, it is vital to generate a dialogical pattern that enables HNRs to demonstrate empathy particularly with people who have psychiatric illnesses [26].

The present-day advanced communication robot systems possess limited functionality in carrying conversations and keeping smooth communication pattern similar to humans, unless this system is connected to a cloud database with distinctive voice assistant services. Using a cloud database with big data capacity complicates information management and security features, increasing risks of data breach and leakage of electronic, sensitive, and confidential data. By installing data security systems, and protective features, HNRs can learn to express more sympathetic behavior over time by undergoing repeated cycles of information processing allowing for secured inputs and outputs of information through the cloud database.

4. The future artificial "brain" for HNRs: associating PsyNACS© with AI systems

What is the brain of a computer? The obligatory answer is the Central Processing Unit (CPU) that performs tons of rapid data processing operations and instructions per second [27]. This is the typical way to define the future artificial "brain" of the HNRs in layman's term. It is metaphorically straightforward to compare human beings with computers wherein human brains and computer processing units function similarly. The CPU or control system is the central nervous system, sensors as the afferent sensory system, and actuators as the efferent motor system [28]. However, this becomes quite difficult and complicated when the task is to describe robot "physiology" and features. For that reason, to successfully characterize the artificial brain of the HNRs in the future, it is critically important to understand the entity that it must emulate – *the professional nurse*.

Amisha et al. described artificial intelligence (AI) as using technology to generate a human-level cognition. In this chapter, the AI does not merely refer to the artificial "brain" of the robot, but rather it characterizes a feature that can understand human language and replicate the behaviors of a professional nurse. To achieve this, AI requires a specialized database like PsyNACS© as well as the capability to communicate verbally and nonverbally. Such ability to control, manage, and operate the HNRs is known as the AI system [29] (**Figure 3**).

HNRs should be able to establish trust and rapport with patients in a similar fashion as a professional human nurse does when fostering a nurse–patient relationship. To have a shared understanding of the patients' life experiences, the HNRs need to understand the patient's illnesses, and treatments. Like nurses, HNRs need to genuinely convey caring to patients and their families through the language of caring.

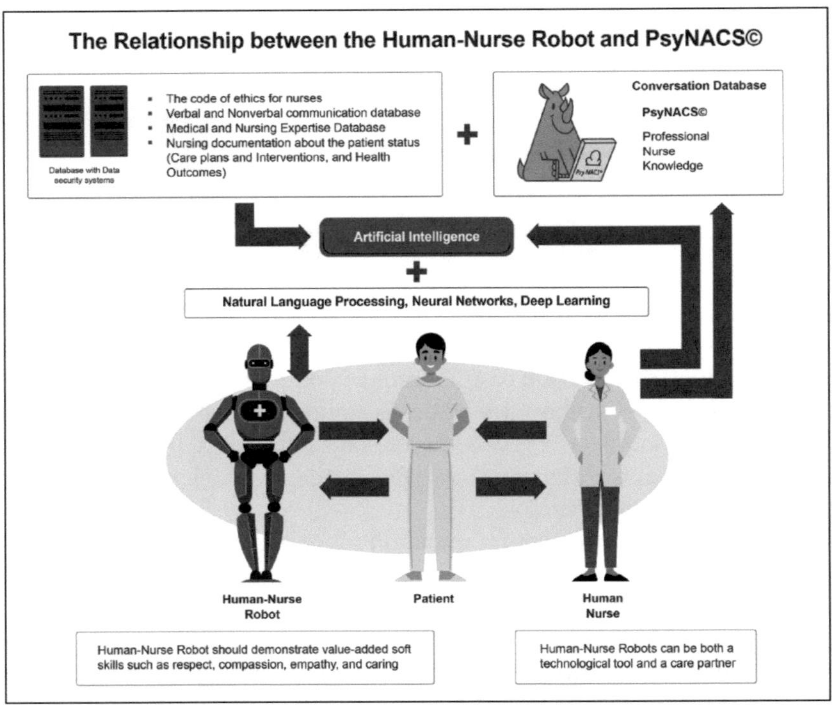

Figure 3.
The relationship between the humanoid-nurse robot and PsyNACS©.

Nurses have self-consciousness that allows them to express their emotions, particularly demonstrating tender loving care without being coached by other people.

Artificial "brains" and artificial consciousness may well be necessary features for HNRs [30] in order to demonstrate initiative and express autonomously without any human inducement or mediation. While nursing care is fundamentally a human-to-human relationship, it becomes a nonhuman-to-human relationship in the case of HNRs [31]. This raises many controversial issues and ethical concerns for patient safety which must be addressed accordingly [24].

If HNRs are to support and care for patients directly, they must hold the same level of comprehensive judgment ability and responsiveness like that of a competent professional nurse who use any of the following processes, such as theory-based nursing care practice, the utility of the traditional nursing process of assessing, diagnosing, planning, implementing, and evaluating, and clinical decision-making, critical thinking, problem-solving, and rapid response and feedback. These processes guide professional practice while emphasizing the individuality of every patient during the practice of professional nursing care. An additional level of intelligence, skillfulness and competence [19] are required in the event that HNRs are assigned to care for patients autonomously or independently.

A successful nurse–patient relationship also relies on effective communication. The future artificial "brain" of HNRs is envisioned to have the capacity to convey a smooth conversation with appropriate patient-centered responses. For HNRs to have such a feature, it entails all the essences of AI such as Natural Language Processing (NLP), neural networks, and deep learning in generating voice contents [32–34]. In addition to verbal content, HNRs need to demonstrate nonverbal communication patterns that are important aspects of effective nurse–patient communications such as eye contact, proxemics, kinesics, expressions, and tone [35–38].

Figure 3 shows the relationship between the robot and PsyNACS© with a conceptual diagram. It may seem therefore that both verbal and nonverbal messages are the life-bloods of successful therapeutic communication in psychiatric nursing. Due to the expected physiological intricacies of the artificial brain, the design and development of HNRs of the future calls for participatory dialog and trans-professional collaboration between healthcare professionals, technology engineers, and care stakeholders. Nursing professionals can provide critical inputs with empirical value at point-of-care. In particular, nurses can contribute to the development of the artificial "brain" for HNRs by sharing their professional knowledge, clinical expertise, care competencies, and nursing documentation that contains relevant and reliable information about the patient status, care plans and interventions, and health outcomes. These information can be organized and amassed using the PsyNACS© framework and database. As a result, a natural conversation *would be* possible between HNRs and humans (e.g., patients and their families) provided that the artificial brain, PsyNACS©, and AI are well-integrated. This allows HNRs of the future to communicate efficiently and respond appropriately and accurately to patients while carefully considering the all-inclusive situation comprising the patient condition, the psychiatric database, and the healthcare environment.

Insofar as robots can be considered as 'mere' artifacts of technological advancement, our trust in and reliance on HNRs must be based on functional and ethical criteria [39]. We can always judge the worth and value of HNRs if their functionalities are approached as means to an end. This teleological approach focuses on the end-result of the HNRs' function that is, whether HNRs have been successful or not in performing tasks. Looking at the outcome itself may overlook the intention of the HNRs. Using a deontological view, we can evaluate HNRs if it is doing the right thing. This also takes into account the goodwill behind the motives and actions of HNRs. Lastly, nursing care is a virtue-driven human experience. We cannot simply assess the HNRs solely based on its obligated and consequential programming. Our evaluation should also consider the value systems in providing quality and safe professional nursing care. As mentioned, HNRs must affirm high-quality, patient-centered, and communication-efficient features. In this light, what can we learn about the value of HNRs in the context of psychiatric nursing care? – With efficiency and wholesome appreciation of being caring entities, *HNRs are more than robots and all the more so than mere tools!* [40].

5. Conclusion

In this chapter, we described the nursing landscape in Japan, PsyNACS©, and HNRs of the future in the context of psychiatric nursing. First, considering the nursing landscape in Japan facilitates a well-defined understanding of the current nursing and healthcare situations to guide the future of psychiatric nursing. Second, a data-driven approach is needed in addressing quality and safety issues in healthcare. We demonstrated how PsyNACS© originated from nursing research, and how we translated it into practice. It allows a secured holistic psychiatric nursing assessment for better care plans and services. The quality of PsyNACS© database content can be enhanced with repeated clinical use. Third, visionary leadership aids in reimagining the future of HNRs to be high-quality, patient-centered, and communication-efficient. Fourth, the artificial "brain" for the HNRs of the future might be incorporated the PsyNACS© database and AI with NLP, neural networks, and deep learning. Collaboration between healthcare professionals, technology engineers, and care stakeholders is essential for the development of HNRs capable of both verbal and nonverbal communication. In summary, integrating PsyNACS© with AI brings HNRs to greater heights – a better quality of nursing care than today.

Acknowledgements

We thank all psychiatric nurses nationwide for their participation in the survey that led to the development of PsyNACS©. This was carried out with the support of JSPS KAKENHI Grant-in-Aid for Young Scientists B (JP16K20819) and the e-Tokushima Promotion Foundation Research Project Grant. We express our gratitude for all of your support.

Conflict of interest

The authors have no conflicts of interest directly relevant to the content of this article.

Author details

Hirokazu Ito[1*], Tetsuya Tanioka[1], Michael Joseph S. Diño[2,3], Irvin L. Ong[2,3] and Rozzano C. Locsin[1,4]

1 Tokushima University, Tokushima, Japan

2 Our Lady of Fatima University, Valenzuela City, Philippines

3 Johns Hopkins University, Baltimore, MD, USA

4 Florida Atlantic University, Christine E. Lynn College of Nursing, FL, USA

*Address all correspondence to: h.itoh@tokushima-u.ac.jp

References

[1] Japanese Nursing Association. Nursing in Japan, Working conditions in Japan [Internet]. 2016. Available from: https://www.nurse.or.jp/jna/english/nursing/employment.html [Accessed: 2020-10-31]

[2] Cabinet Office, Government of Japan. The Situation and Future of the Elderly, White Paper on Aging Society (Overall Version) (In Japanese) [Internet]. 2019. Available from: https://www8.cao.go.jp/kourei/whitepaper/w-2019/zenbun/pdf/1s1s_01.pdf [Accessed: 2020-10-31]

[3] Ministry of Health, Labor and Welfare. Changes in Labor Force: Annual Report on Health, Labor and Welfare. (In Japanese) [Internet]. 2017. Available from: https://www.mhlw.go.jp/wp/hakusyo/kousei/17-2/dl/01.pdf [Accessed: 2020-10-31]

[4] Robinson H, Macdnald B, Sroadbent E: The Role of Healthcare Robots for Older People at Home: A Review. International Journal of Social Robotics. 2014;6:575-591. DOI: 10.1007/s12369-014-0242-2

[5] Ministry of Health, Labor and Welfare. Appendix 5-3 Welfare Tools Related to Priority Areas in the Use of Robot Technology for Nursing Care. (In Japanese) [Internet]. Available from: https://www.mhlw.go.jp/file/05-Shingikai-12301000-Roukenkyoku-Soumuka/0000094783.pdf [Accessed: 2020-10-31]

[6] Ministry of Health, Labor and Welfare. Annual Reports on Health and Welfare 1998-1999 Part 1 Social Security and National Life, Chapter 4 Section 2. Review of Health and Welfare Measures for People with Disabilities [Internet]. Available from: https://www.mhlw.go.jp/english/wp/wp-hw/vol1/p2c4s2.html [Accessed: 2020-10-31]

[7] Tanioka T, Chiba S, Onishi Y, et al.: Factors associated with discharge of long-term inpatients with schizophrenia in Japan: a retrospective study. Issues Ment Health Nurs. 2013;34(4):256-264. DOI:10.3109/01612840.2012.742602

[8] Corrigan PW, Green A, Lundin R, Kubiak MA, Penn DL: Familiarity with and Social Distance from People Who Have Serious Mental Illness. Psychiatric Services. 2001;52:953-958. DOI: 10.1176/appi.ps.52.7.953

[9] Crisp AH, Gelder MG, Rix S, Meltzer HI, Rowlands OJ: Stigmatisation of People with Mental Illnesses. The British Journal of Psychiatry. 2000;177:4-7. DOI: 10.1192/bjp.177.1.4

[10] World Health Organization. Disability and Health. [Internet]. 2020. Available from: http://www.who.int/mediacentre/factsheets/fs352/en/ [Accessed: 2020-12-1]

[11] Jobe TH, Harrow M: Long-Term Outcome of Patients with Schizophrenia, A Review. The Canadian Journal of Psychiatry. 2005;50:892-900. DOI: 10.1177/070674370505001403

[12] Gastal FL, Andreoli SB, Quintana MI, Gameiroa MA, Leitea SO, McGrathc J: Predicting the Revolving Door Phenomenon among Patients with Schizophrenic, Affective Disorders and Non-Organic Psychoses. Revista de Saúde Pública. 2000;34:280-285. DOI: 10.1590/S0034-89102000000300011

[13] Frick U, Frick H, Langguth B, Landgrebe M, Hübner-Liebermann B, Hajak G: The Revolving Door Phenomenon Revisited. Time to Readmission in 17′415 Patients with 37′697 Hospitalisations at a German Psychiatric Hospital. PLoS One. 2013;8:e75612. DOI: 10.1371/journal.pone.0075612

[14] Cerejeira J, Lagarto L, Mukaetova-Ladinska EB: Behavioral and psychological symptoms of dementia. Front. Neurol. 2012;3(73):1-21. DOI: 10.3389/fneur.2012.00073

[15] Ito H, Miyagawa M, Ozawa K, Tanioka T, Yasuhara Y, Kataoka M, King B, Tomotake M, Locsin R: The Development of the Japanese Psychiatric Nursing Assessment Classification System (PsyNACS©). Open Journal of Psychiatry. 2015;6(1):20-33.

[16] Ito H, Miyagawa M, Ozawa K, Tanioka T, Yasuhara Y, Kataoka M, King B, Tomotake M, Kuwamura Y, Locsin RC: Prioritizing Patient Assessment Data (PAD) Using the Japanese Psychiatric Nursing Assessment Classification System (PsyNACS)©. Open Journal of Psychiatry. 2016;6(3):218-227. DOI: 10.4236/ojpsych.2016.63027

[17] Vaidya S, Ambad P, Bhosle S: Industry 4.0 – A Glimpse. Procedia Manufacturing. 2018;20:233-238. DOI: 10.1016/j.promfg.2018.02.034

[18] Cabinet Office, Government of Japan. What is Society 5.0? [Internet]. Available from: https://www8.cao.go.jp/cstp/english/society5_0/index.html [Accessed: 2020-10-31]

[19] Tanioka T, Osaka K, Locsin R, Yasuhara Y, Ito H: Recommended Design and Direction of Development for Humanoid Nursing Robots Perspective from Nursing Researchers. Intelligent Control and Automation. 2017;8:96-110. DOI: 10.4236/ica.2017.82008.

[20] Tanioka T, Smith MC, Osaka K, Zhao Y: Framing the Development of Humanoid Healthcare Robots in Caring Science. International Journal for Human Caring. 2019;23(2):112-120.

[21] Schoenhofer SO, Van Wynsberghe A, Boykin A: Engaging robots as nursing partners-in-caring: Nursing As Caring meets Care-Centered Value Sensitive Design. International Journal for Human Caring. 2019;23:157-167.

[22] Boykin A, Schoenhofer SO. Nursing as caring: A model for transforming practice. 2nd ed. Sudbury, MA: Jones and Bartlett; 2001. 71p.

[23] Van Wynsberghe A: Designing robots for care: Care centered value-sensitive design. Science and Engineering Ethics, 2013;19(2):407-433. DOI: 10.1007/s11948-011-9343-6

[24] Yasuhara Y, Tanioka R, Tanioka T, Ito H, Tsujigami Y: Ethico-Legal Issues with Humanoid Caring Robots and Older Adults in Japan. International Journal for Human Caring. 2019;23(2):141-148.

[25] Yasuhara Y, Tanioka T, Kai Y, et al.: Potential legal issues when caring healthcare robot with communication in caring functions are used for older adult care. Enfermería Clínica. 2020;30(Suppl 1):54-59. DOI: 10.1016/j.enfcli.2019.09.024

[26] Miyagawa M, Yasuhara Y, Tanioka T, Locsin R, Kongsuwan W, Catangui E, Matsumoto K: The Optimization of Humanoid Robot's Dialog in Improving Communication between Humanoid Robot and Older Adults. Intelligent Control and Automation, 2019;10:118-127. DOI: 10.4236/ica.2019.103008

[27] Indrajit IK. and Alam A: Computer hardware for radiologists: Part I. Indian J Radiol Imaging. 2010; 20(3): 162-167. DOI: 10.4103/0971-3026.69346

[28] Schwartz BA: Movement: How the Brain Communicates with the World. Cell. 2016; 164(6): 1122-1135. DOI: 10.1016/j.cell.2016.02.038

[29] Amisha, Malik P, Pathania M, Rathaur VK.: Overview of artificial intelligence in medicine. J Family

Med Prim Care. 2019;8(7):2328-2331. doi:10.4103/jfmpc.jfmpc_440_19

[30] Zhao Y, Kai Y, Tanioka T: Discovering Humanoid Robot Consciousness Within Caring Science. International Journal for Human Caring. 2019;23(2):121-131.

[31] Tanioka T: The Development of the Transactive Relationship Theory of Nursing (TRETON): A Nursing Engagement Model for Persons and Humanoid Nursing Robots. International Journal of Nursing & Clinical Practices 2017;4:223. DOI: https://doi.org/10.15344/2394-4978/2017/223

[32] Settles B. Active learning literature survey. In: Computer Sciences Technical Reports. University of Wisconsin-Madison Department of Computer Sciences. 2009. 34p. http://www.burrsettles.com/pub/settles.activelearning.pdf

[33] Reyes O, Morell C, Ventura S: Effective active learning strategy for multi-label learning. Neurocomputing. 2018;273:494-508. DOI: https://doi.org/10.1016/j.neucom.2017.08.001

[34] Zhang X, Cheng J, Xu C, Lu H, Ma S. Multi-view multi-label active learning for image classification. In: 2009 IEEE International Conference on Multimedia and Expo; 27-30 October 2014; New York. New York: IEEE;2009. p. 258-261. DOI: 10.1109/ICIP.2014.7026058

[35] Mavridis N: A review of verbal and non-verbal human–robot interactive communication. Robotics and Autonomous Systems. 2015;63(1):22-35. DOI: 10.1016/j.robot.2014.09.031

[36] Caris-Verhallen WMCM, Kerkstra A, Bensing JM, Grypdonck M: Effects of video interaction analysis training on nurse-patient communication in the care of the elderly. Patient Education and Counseling. 2000;39(1):91-103. DOI: 10.1016/S0738-3991(99)00094-4

[37] Davies P: Non-verbal communication with patients. British Journal of Nursing. 1994;3(5):220-223. DOI: 10.12968/bjon.1994.3.5.220

[38] Khalifa EM, Faddal H. Impacts of Using Paralanguage on Teaching and Learning English Language to Convey Effective Meaning. Studies in English Language Teaching. 2017;5(2):295-306. DOI: 10.22158/selt.v5n2p295

[39] Cross ES, Hortensius R, Wykowska A: From social brains to social robots: applying neurocognitive insights to human–robot interaction. Phil. Trans. R. Soc. 2019;374(1771):20180024. DOI: 10.1098/rstb.2018.0024

[40] Locsin R, Ito H, Tanioka T, Yasuhara Y, Osaka K, Schoenhofer SO: Humanoid Nurse Robots as Caring Entities: A Revolutionary Probability?. International Journal of Studies in Nursing, 2018; **3**(2): 146-154. DOI: 10.20849/ijsn.v3i2.456

Chapter 8

Expectations and Ethical Dilemmas Concerning Healthcare Communication Robots in Healthcare Settings: A Nurse's Perspective

Yuko Yasuhara

Abstract

This chapter describes expectations and ethical dilemmas concerning healthcare communication robots (HCRs) from a nurse's perspective. Ethical dilemmas in nursing settings are wide-ranging. When HCRs are introduced to long-term facilities and hospitals for patient communication, new kinds of ethical dilemmas may arise. Using interviews with healthcare providers, I examined the potential ethical dilemmas concerning the development and introduction of HCRs that may interact with older adults. This analysis was based on four primary issues from the nurses' perspective. Since HCRs will be used in healthcare settings, it is important to protect patient rights and maintain their safety. To this end, discussion and collaboration with an interdisciplinary team is crucial to the process of developing these robots for use among patients.

Keywords: healthcare communication robots, ethical dilemmas, nurse's perspective

1. Introduction

Japan's declining birthrate and aging population are becoming increasingly serious issues. Indeed, the shrinkage of the working population continues unabated [1]. The Ministry of Health, Labour and Welfare [2] reported a future shortage of anywhere from 60,000 to 270,000 nurses in 2025. This shortage might make it difficult to provide sufficient patient care, especially for older adults who need long-term care [3].

Beyond this, as of early November 2020, coronavirus disease 2019 (COVID-19) became a major threat to global public health. Globally, the number of patients with COVID-19 is approximately 52 million [4]. Notably, COVID-19 is caused by the SARS-CoV-2 virus, which spreads among people, mainly when an infected person is in close contact with others [5]. Significantly, many COVID-19 clusters have been reported in clinical settings, including long-term facilities.

In Japan, the Ministry of Health, Labor and Welfare [6] recommends that people employ basic strategies to prevent the spread of infectious diseases, including COVID-19. These include hand washing, proper cough etiquette, wearing a mask, and avoiding group gatherings in poorly ventilated spaces. Although potential vaccines are under development, it remains necessary to make lifestyle changes that extend to human interaction, recognizing the possibility that new infectious diseases may gain prevalence in the future.

Robots are attracting attention as a countermeasure for such serious situations. Of the various forms of human interaction, communication with others is important as it helps improve the quality of life (QOL) and sociality of older adults and patients with dementia. Accordingly, healthcare communication robots (HCRs) have the potential to support the needs of patient dialog as an alternative to healthcare providers, thereby preventing infections and addressing staff shortage situations.

Using HCRs for patient care is a collaborative process that requires not only engineers but also healthcare providers, such as nurses, who have a mandate to protect patient rights and maintain safety. Indeed, it is necessary to consider potential issues that may arise from this development. Thus, this paper discusses expectations and ethical dilemmas in relation to HCRs from the perspective of nurses.

2. Expectations from healthcare communication robots

Communication with others is important because it is satisfying and fosters a sense of connection. Especially, conversation with others achieves mutual understanding through shared experiences and feelings. However, in Japan, community relationship networks are becoming degraded by the progressively aging society and the trend of nuclear families, which have become serious local problems. Particularly among older adults who have lived alone or had physical functional disorder, social activity and conversation with others tend to decrease.

Notably, long-term facilities have seen a rise in dementia patients, and the behavioral and psychological symptoms of dementia (BPSD) may cause irritability and restlessness among patients [7]. When nurses care for older adults and patients with dementia, it is important that they take time to listen to them to provide appropriate, high-quality care in a way that suits the patient [8].

However, the staffing of nurses in long-term facilities and nursing homes for older adults is lower than in acute care hospitals [9]. Due to this shortage of healthcare providers, it might be burdensome for staff to take sufficient time for dialog with older adults [10, 11].

Clearly, the quality of care for older adults may be suffering because of labor shortages, especially in long-term care settings. This quality of care may be expected to improve when healthcare workers have HCRs as partners. Moreover, HCRs may also provide patients with the opportunity to talk, even in situations where an infectious disease such as COVID-19 is concerned.

The Japanese government has already supported the introduction of HCRs to facilities for the elderly (such as nursing homes) as well as healthcare facilities [12] and hospitals [13]. While HCRs are still being developed and introduced in certain facilities, there are no HCRs specialized for older adults and patients with dementia [14]. Hence, it is necessary to improve the application that enables dialog with members of these demographics and to enhance the safety and features of the robots [15].

The development of HCRs capable of dialog and therapeutic communication is a future goal. Here, "dialog" is not just a conversation, but the recognition and respect for each other's values and establishing a relationship of trust.

This speaks to the larger need for the development of HCRs that can interact with the elderly, increase conversation opportunities for them, satisfy their desire for approval, maintain their sociality and sense of purpose, and improve their QOL. Furthermore, by collecting information from the cloud database of these robots, healthcare providers may be able to determine whether urgent or immediate care is necessary, allowing them to listen to the patients more intensely.

The acute care field is marked by the responsibility to care for patients suffering from threatening infectious diseases such as COVID-19. The risk of infection is very high for medical staff [16], who must find a way to take care of patients within the boundaries of time constraints, while also striving to prevent getting infected. Unsurprisingly, most medical staff find it difficult to take enough time to listen to patients' feelings, particularly when they are fighting the fear of COVID-19 infection [17, 18]. Thus, patients with COVID-19 may lose the opportunity to express themselves because they have limited time to talk to their medical staff and limited visits with family and friends.

Traditional (human) nurses are accustomed to listening to a patient's voice. However, in an emergency, HCRs may be able to note a patient's anxiety and complaints and provide them with appropriate care in response. If the HCR can be linked with information from thermography and electronic medical records, it will also be possible to observe simple physical conditions among patients. Thus, the HCR may also serve as an alternative to care supporters for people who have been in shelters for long periods due to large earthquakes, etc.

3. Ethics required of nurses

As recent years have seen the rapid development of robots and artificial intelligence (AI), ethical codes and guidelines have been issued by related academic societies largely in the engineering field [19, 20]. Ethical studies concerning AI and robots are also underway. UK-RAS network describes that the ethical concerns raised by robotics and autonomous systems (RAS) depend on their capabilities and domain of usage of Robotics, there are ethical issues such as Bias, Deception, Employment, Opacity, Safety, Oversight, and Privacy [21]. Of course, ethics are crucial to healthcare because healthcare workers must recognize dilemmas: using good judgment to make decisions informed by their values but also governed by the law.

A nurse, a type of healthcare provider, is a person who engages in providing care to persons with injuries and/or illnesses, and/or postpartum women, and/or assists in the provision of medical treatment under the license of the Ministry of Health, Labour and Welfare (Article 5 of the Act on Public Health Nurses, Midwives, and Nurses). Based on the Nursing Code of Ethics of the International Council of Nurses (ICN) [22], and the Japanese Nursing Association (JNA) [23], nurses are required to provide care while respecting human life, dignity, and rights according to the law.

However, just as patients are unique and vary in age and condition, nurses have their own cultural, religious, moral, and professional values. Thus, there are often conflicting values, disagreements, and ethical conflicts in nursing settings.

Ethical dilemmas in nursing settings are far-reaching. From time to time, nurses make ethical decisions by taking a variety of information into account to determine the best choice for the patient. Nurses can take appropriate actions when faced with an ethical dilemma by understanding and applying ethical guidelines such as the American Nurses Association's Code of Ethics [24], the ICN Code of Ethics for Nurses [22], and the JNA Code of Ethics [23].

In Japan, decisions about ethical dilemmas are informed by the six principles of ethics (Beneficence, Non-maleficence, Autonomy, Veracity, Justice, and Fidelity)

Beneficence: Actions that consider the welfare of others and include attributes like kindness and charity.
Nonmaleficence: Actions that prevent or inflict minimal harm to others.
Autonomy: Recognizing the individual's right to self-determination and decision-making.
Veracity: Interacting with others in a truthful, trustworthy, and accurate manner.
Justice: Treating others with fairness and with equal degree of respect and concern.
Fidelity: Being loyal and faithful to patients who trust the nurse.

Table 1.
Six principles of ethics [25–27].

(**Table 1**) [25, 26]. These principles are familiar to nurses. Even after making ethical decisions, nurses reflect on those decisions and strive to increase their ethical sensitivity daily.

When HCRs are introduced to long-term facilities and hospitals, different ethical dilemmas might occur.

If the HCRs, in the near future, can use dialog to make autonomous decisions regarding patients, and serve to replace a human nurse, relevant ethical discussions must precede this change. For instance, one would logically consider the questions of whether HCRs can have a sense of ethics like human nurses, and whether the former can make ethical decisions in the midst of ethical conflicts within nursing settings.

4. A nurse's perspective on ethical dilemmas regarding healthcare communication robots

Our research currently uses the humanoid robot, Pepper (SoftBank Robotics Corp.) [28], in a long-term facility to develop an application for healthcare robots that can communicate with older adults based on principles of care. It also seeks to evaluate a program that can be run in a clinical context (developed by the Xing Company). However, in the implementation of this strategy, the communication function of Pepper's application has proven deficient.

It is important to understand the present HCRs' competency as well as other factors that may enhance this application, making it suitable for use among older adults. To explore HCR-related issues in healthcare settings, we interviewed five healthcare providers (nurses, caregivers, and physiotherapists) at three facilities about current usage issues with Pepper. From these results, I examined ethical dilemmas from the nurse's perspective concerning the development and introduction of HCRs that can interact with older adults. This analysis was based on four issues: burden on staff and insufficient support system, inadequate communication function, leakage of personal information and violation of right to privacy, and guaranteeing the safety and security of HCRs.

4.1 Burden on staff and insufficient support system

The complexity of the robot's operation, the ambiguity of the HCR support system, and the burden of preparation and cleanup of HCRs are some of the issues faced by the staff while working with HCRs. Pepper weighs approximately 30 kg (around 66 lbs.), stands 120 cm (approximately 47 inches) tall [28], and requires extra staff to prepare it for use and clean it. In addition, there are other issues related to its operational complexity and unclear support system (e.g., where to check when the robot freezes). These issues sometimes occur due to the application's up-data. In many cases, a specific healthcare staff member accustomed to handling such equipment is in charge of making the introduction, placing additional burden on that staff member. At such times, staff support is required to facilitate interactions and conversations between humans and robots [29].

When HCRs are used in healthcare settings, it is important to avoid increasing the human burden and preventing the traditional nurse from being deprived of time to care for the patient. This is related to the ethical principle of justice. Nurses must decide the just or fair allocation of healthcare resources [25, 26]. With the introduction of robots, the principles of beneficence (providing good nursing to all patients), non-maleficence (avoiding harm caused from using HCRs), and justice (providing proper and fair nursing to all patients) should not come at the cost of staff conflict. Undue burden placed upon nurses, such as the aforementioned HCR handling and use requirements, may incline nurses to put an end to the introduction of robots in healthcare settings as they cannot provide adequate care and ensure the patient's safety. Indeed, convenience (which includes appropriate sizing) and generous support are key for HCR use. It is also necessary to have functions that can be used by medical professionals who are not well versed in robotics and/or engineering.

4.2 Inadequate communication function

A human nurse naturally changes the manner (speed, volume, delivery, tone) and content of their speech depending on the patient, the nurse's personal experience, and various other factors. Conversely, the current HCRs cannot change how they talk to patients. Thus, older adults and patients with dementia may give up the conversation, feel discouraged, and/or experience negative emotions because the timing of HCRs' utterances and the content of the response may be insufficient and the conversations may be unengaging. This has implications for the ethical principle of non-maleficence.

The challenge here is to set the goals for the HCRs' dialog function to include the examination of word choice (including the determination of inappropriate words). Clearly, the dialog function will rapidly improve in the future. However, traditional nurses are currently better placed to provide care to patients based on nursing ethics and while exercising professional responsibility.

Even during the clinical trials for HCR development, nurses must protect patients' rights. Patients should not be harmed; they should not experience negative feelings or feel discouraged by HCRs (the principle of non-maleficence). Nurses should ensure that patients receive the best care from HCRs and human nurses (the principle of beneficence). Furthermore, it is particularly important to solicit patients' opinions concerning their willingness or desire to interact with the HCRs (principle of autonomy); they should be permitted the personal liberty to determine their own decisions on whether to receive care from HCRs [25, 26]. Nurses give top priority to the safety of the subject and thereby play an advocacy role. Therefore, if patient rights and their ethical principles are violated, nurses may need to halt the promotion of robot development.

4.3 Leakage of personal information and violation of right to privacy

The third issue involves the collection of patient information stored in the cloud server or body of HCRs, and how this information is managed. Indeed, HCRs need to store information to a cloud server for improved functioning. A cloud server allows for information input from various sources, along with simultaneous compilation and analysis [30]. This is significant, as there is a lot of information in the dialog between patients and HCRs.

The guidelines regarding AI and robots have included effective policies such as protection and promotion of human rights, safety, and privacy [19, 20, 31]. Nevertheless, in the near future, when HCRs use the cloud server to store big data collected from their patients, an information leakage accident may occur [32]. This issue could, for instance, arise due to some malfunction during the development stage.

The right to privacy does not have a legal basis in Japan. However, the right to privacy is recognized under the law of precedent as part of the pursuit of happiness referred to in Article 13 of the Constitution. In addition, personal information, in principle, cannot be provided to a third party (Article 23), except in cases where the allowance is based on laws and regulations (Article 23–1).

Nurses also have a duty to protect patients' privacy as a component of patient care (Article 42–2 of the Act on Public Health Nurses, Midwives, and Nurses). As stated in the code of nurse ethics, "Nurses should honor confidentiality and strive for the protection of personal information, while using appropriate discretion in the sharing of this information" [32]. Hence, it is important to safeguard against personal information leakage from HCRs or iCloud servers (the principles of fidelity, and non-maleficence).

4.4 Guaranteeing the security/safety of healthcare communication robots

The fourth issue is the need to ensure the safety of interactive robots. In healthcare settings, there are hazardous things that might result in daily medical accidents or incidents. A medical accident involving a nurse may happen while providing nursing care or while assisting medical treatment that involves medical interventions [33]. Healthcare institutions continue to improve their policies and framework to secure organization-wide safety [34]. Nurses consistently make patient safety a top priority (the principles of non-maleficence: avoiding harm caused by HCRs, and beneficence: providing better nursing to all patients). This consideration entails predicting potentially dangerous patient behavior and performing other forms of safety and risk management (the principle of non-maleficence).

Presently, there are no reported medical accidents due to the use of HCRs. Unless there is a guarantee that accidents due to patient falls or contact will not occur, and that the safety of nurses and medical staff will be ensured, the introduction of HCRs should not be viewed passively.

For instance, we must consider whether HCRs that can interact with older adults and patients with dementia need a self-propelled function and/or humanoid figures, and whether these things would enhance patient safety. Moreover, different cases must be studied along with the safety-related responsibilities they present.

5. Conclusion

This chapter discusses expectations and ethical dilemmas concerning the use of HCRs that will interact with patients in medical and welfare settings in the future. These considerations have been made from the nurses' perspective.

Conversation with others is important to human beings. However, appropriate reactions and responses are complex, not just for HCRs, but also for traditional nurses. This means that, HCRs require improved functions, including specifications concerning appropriate listening practices, conversation, behavior, etc.

Furthermore, nurses must continue to protect the rights and safety of patients in all instances and at all times. Thus, HCRs should not be allowed to infringe on these principles in healthcare settings.

In the future, HCRs may serve as patient interlocutors. Their conversation program may include AI with an interactive or transactive dialog function and the capacity to make decisions concerning ethical conflicts. To this end, discussion and collaboration with an interdisciplinary team is crucial to the process of developing these robots for use among patients.

Acknowledgements

This work was partially supported by JSPS KAKENHI Grant Number JP19K10735. Part of this work was presented to the Japan Society of Mechanical Engineers in 2020.

Conflicts of interest

The authors declare no conflicts of interest.

Author details

Yuko Yasuhara
Department of Nursing Outcome Management, Institute of Biomedical Sciences, Tokushima University, Graduate School, Tokushima, Japan

*Address all correspondence to: yaushara@tokushima-u.ac.jp

References

[1] Ministry of Health, Labour and Welfare. Long-Term Care Insurance System of Japan [Internet]. Available from: http://www.mhlw.go.jp/english/policy/care-welfare/care-welfare-elderly/dl/ltcisj_e.pdf. Published November 2016. [Accessed: 20-11-25]

[2] Ministry of Health, Labour and Welfare. Kangosyokuinn jyuyou bunnkakai, tyukan torimatome (The nursing staff supply and demand committee, an interim report) (in Japanese) Available from: https://www.mhlw.go.jp/content/12601000/000567798.pdf. Published November 18, 2019. [Accessed: 2020-11-20]

[3] Japanese Nursing Association. Nursing in Japan. Available from: https://www.nurse.or.jp/jna/english/nursing/employment.html. [Accessed 2020-11-20]

[4] World Health Organization. WHO Coronavirus Disease (COVID-19) Dashboard. World Health Organization [Internet]. Available from: https://covid19.who.int/. Published November 18, 2020. [Accessed: 2020-11-10]

[5] World Health Organization. Coronavirus disease (COVID-19): How is it transmitted? [Internet]. Available from: https://www.who.int/news-room/q-a-detail/q-a-how-is-covid-19-transmitted. Published July 9, 2020. [Accessed: 2020-11-18]

[6] Ministry of Health, Labour and Welfare. Preventing COVID-19 (novel coronavirus) and stopping its spread. Covid-19 info [Internet]. Available from: https://www.covid19-info.jp/covid-19-en.html#contents01. [Accessed: 2020-11-10]

[7] Cerejeira J, Lagarto L, Mukaetova-Ladinska EB. Behavioral and psychological symptoms of dementia. Front. Neurol. 2012;3(73):1-21. DOI: https://doi.org/10.3389/fneur.2012.00073

[8] Tible OP, Riese F, Savaskan E, von Gunten A. Best practice in the management of behavioral and psychological symptoms of dementia. Ther Adv Neurol Diso. 2017;10(8):297-309.

[9] Japanese Nursing Association. Nursing for the older people in Japan [Internet]. Available from: https://www.nurse.or.jp/jna/english/pdf/info-02.pdf. [Accessed: 2020-11-18]

[10] Oba H. Assessment of cognitive functions by communication with people having dementia, Journal of Care and Behavioral Science for the Elderly. 2018;23:2-10. DOI: https://doi.org/10.24777/jsbse.23.0_2

[11] Mallidou AA, Cummings GG, Schalm C, Estabrooks CA. Health care aides use of time in a residential long-term care unit: A time and motion study. Int J Nurs Stud. 2013;50:1229-1239.

[12] Ministry of Economy, Trade and Industry. Revision of the Priority Areas to Which Robot Technology is to be Introduced in Nursing Care [Internet]. Available from: https://www.meti.go.jp/english/press/2017/1012_002.html. Published October 2017. [Accessed: 2020-10-18]

[13] Šabanović S. Inventing Japan's 'robotics culture': The repeated assembly of science, technology, and culture in social robotics. Soc Stud Sci. 2014;44(3):342-367. DOI: https://doi.org/10.1177/0306312713509704

[14] Miyagawa M, Yasuhara Y, Tanioka T, Locsin RC, Kongsuwan W, Catangui E, Matsumoto K. The Optimization of Humanoid Robots Dialog in Improving Communication between

Humanoid Robot and Older Adults. Intelligent Control and Automation. 2019;10(3):118-127. DOI: https://doi.org/10.4236/ica.2019.103008

[15] Yasuhara Y, Tanioka T, Kai Y, TsujigamiY, Uematsu K, Dino JM, Locsin RC, Schoenhofer SO. Potential legal issues when caring healthcare robot with communication in caring functions are used for older adult care. Enferm Clin. 2020;30(1):54-59. DOI: 10.1016/j.enfcli.2019.09.024

[16] Gheysarzadeh A, Sadeghifard N, Safari M, Balavandi F, Falahi S, Kenarkoohi A Tavan H. Report of five nurses infected with severe acute respiratory syndrome coronavirus 2 during patient care: case series. New Microbes New Infect. 2020;36:1-3. DOI: https://doi.org/10.1016/j.nmni.2020.100694

[17] Labrague LJ, de Los Santos JAA. Fear of COVID-19, psychological distress, work satisfaction, and turnover intention among frontline nurses. J Nurs Manag. 2020;10:1-8. DOI: 10.1111/JONM.13168

[18] Khattak SR, Saeed I, Rahman SU, Fayaz M. Impact of Fear of COVID-19 Pandemic on the Mental Health of Nurses in Pakistan. Journal of Loss and Trauma, International Perspectives on Stress & Coping. 2020, 1-15. DOI: https://doi.org/10.1080/15325024.2020.1814580

[19] Institute of Electrical and Electronic Engineers. Ethically Aligned Design, A Vision for Prioritizing Human Wellbeing with Artificial Intelligence and Autonomous Systems (AI/AS). Global Initiative for Ethical Considerations in Artificial Intelligence and Autonomous Systems. Available from: https://standards.ieee.org/content/dam/ieee-standards/standards/web/documents/other/ead_v1.pdf. Published December 13, 2016. [Accessed: 2020-11-20]

[20] The Ethics Committee of the Japanese Society for Artificial Intelligence (JSAI). Ethical Guidelines [Internet]. Available from: http://ai-elsi.org/wp-content/uploads/2017/05/JSAI-Ethical-Guidelines-1.pdf. Accessed Nov 23, 2020.

[21] UK RAS Network, Ethical Issues for Robotics and Autonomous Systems: 2019. Available from: https://www.ukras.org/wp-content/uploads/2019/07/UK_RAS_AI_ethics_web_72.pdf.

[22] International Council of Nurses. The ICN Code of Ethics for Nurses: Revised 2012. Available from: https://www.icn.ch/sites/default/files/inline-files/2012_ICN_Codeofethicsfornurses_%20eng.pdf. Published month date, year. [Accessed: 2020-11-25]

[23] Japanese Nursing Association, JNA Code of Ethics for Nurses. Available from: https://www.nurse.or.jp/jna/english/activities/pdf/ethics2003.pdf. [Accessed: 2020-11-22]

[24] American Nurses Association. Ethics and Human Rights. Available from: https://www.nursingworld.org/practice-policy/nursing-excellence/ethics/. [Accessed 2020-11-22]

[25] Fry ST, Veatch RM, Case studies in nursing ethics, 3rd ed. Boston: Jones & Barlett.

[26] Fry ST, Johnstone MJ. Ethics in Nursing Practice: A guide to ethical decision making. 3rd ed. Geneva, Switzerland: International Council of Nurses (ICN); 2008.

[27] Lockwood W. Nursing Ethics. Available from: https://www.rn.org/courses/coursematerial-10026.pdf. Published April 2020. [Accessed: 2020-11-18].

[28] SoftBank Robotics (n.d.) Pepper. Available from: https://www.

softbankrobotics.com/emea/en/pepper. [Accessed: Y2020-12-24]

[29] Tanioka T, Yasuhara Y, Dino Joseph S. Michael, Kai Y, Locsin RC, Savina S. Disruptive Engagements with Technologies, Robotics, and Caring: Advancing the Transactive Relationship Theory of Nursing. Nurs Adm Q. 2019;43(4):313-321.

[30] Zheng Z, Wang P, Liu J, Sun S. Real-time big data processing framework: Challenges and solutions. Appl. Math. Inf. Sci. 2015;9(6):3169-3190. Retrieved October 29, 2018 from http://www.naturalspublishing.com/files/published/v6910010rnl56m.pdf

[31] Institute of Electrical and Electronic Engineers. The IEEE Global Initiative on Ethics of Autonomous and Intelligent Systems [Internet]. Available from: https://standards.ieee.org/content/dam/ieee-standards/standards/web/documents/other/ead1e-introduction.pdf?utm_medium=undefined&utm_source=undefined&utm_campaign=undefined&utm_content=undefined&utm_term=undefined. [Accessed 2020-11-23]

[32] Yasuhara Y, Tanioka R, Tanioka T, Ito H, Tsujigami Y. Ethico-Legal Issues with Humanoid Caring Robots and Older Adults in Japan. Int J Hum Caring. 2019;23(2):141-148.

[33] Japanese Nursing Association, Nursing and Patient Safety in Japan, 2. Nursing and patient safety [Internet]. Available from: https://www.nurse.or.jp/jna/english/pdf/safety-02.pdf [Accessed: 2020-12-5].

[34] Japanese Nursing Association, Nursing and Patient Safety in Japan, 1. Nursing and patient safety [Internet]. Available from: https://www.nurse.or.jp/jna/english/pdf/safety-01.pdf [Accessed: 2020-12-5].